THE KALEIDOSCOPE OF SCIENCE

BOSTON STUDIES IN THE PHILOSOPHY OF SCIENCE

EDITED BY ROBERT S. COHEN AND MARX W. WARTOFSKY

VOLUME 94

THE KALEIDOSCOPE
OF SCIENCE

*The Israel Colloquium: Studies in
History, Philosophy, and Sociology of Science*

Volume 1

Edited by

EDNA ULLMANN-MARGALIT

SPRINGER-SCIENCE+BUSINESS MEDIA, B.V.

Library of Congress Cataloging in Publication Data
Main entry under title:

The Kaleidoscope of science.

    (Boston studies in the philosophy of science ; 94)
    Includes index.
    1. Science–Philosophy. 2. Science–History.
I. Ullmann-Margalit, Edna. II. Series.
Q174.B67   vol. 94   001'.01s [501]   85–28167
[Q175]
ISBN 978-90-277-2159-4   ISBN 978-94-009-5496-0 (eBook)
DOI 10.1007/ 978-94-009-5496-0

Prepared in cooperation with Mrs. Esther Shashar, executive editor,
The Van Leer Jerusalem Institute.

THE ISRAEL COLLOQUIUM: STUDIES IN HISTORY, PHILOSOPHY, AND SOCIOLOGY OF SCIENCE

The Israel Colloquium for the History, Philosophy, and Sociology of Science was established in the academic year 1981–82. It offers, annually, a series of public lectures, alternately in Jerusalem and Tel-Aviv. It is sponsored and directed jointly by three bodies: The Center for the History and Philosophy of Science, Technology, and Medicine of The Hebrew University of Jerusalem; The Institute for the History and Philosophy of Science and Ideas of Tel-Aviv University; and The Van Leer Jerusalem Institute. It has an advisory board representing all the institutions of higher learning in Israel; also, it collaborates with R. S. Cohen and M. Wartofsky of the Boston Colloquium for the Philosophy of Science.

EDITORIAL NOTE

On the Proceedings of the Israel Colloquium for the History, Philosophy and Sociology of Science:

In recent years it has become evident that, in addition to serious and competent disciplinary work on the specifics of the History of Science, the Philosophy of Science and the Sociology of Science, there is now a growing need to develop a problem-oriented approach which no longer distinguishes between these three specialties in a cut and dried way. Since the time has come for such an approach, the institutional tools should be provided.

A way to do so would be to write – and to encourage others to write – almost 19th century style, problem-oriented overviews and to devote journals and workshops to that as a systematic approach. Another way is more eclectic, namely to organize colloquia and to publish good papers stemming from these, without attempting to organize the papers under the separate rubrics of History or Philosophy or Sociology of Science; and moreover to consider it natural that any fundamental issue of the foundations of the sciences, or their place in a culture and the way they are institutionalized in the societal web, is still our concern, no matter whether we are a professional scientist, historian or philosopher who deals with the problem.

The Boston Colloquium for the Philosophy of Science at Boston University, which celebrated its 25th anniversary this year, represented this approach under the general umbrella of Philosophy of Science. The Israel Colloquium, which is now in its fifth year, started as an offspring of the Boston Colloquium with the advice and encouragement of the founders and organizers of the Boston Colloquium, and has chosen explicitly to devote itself to History, Philosophy and Sociology of . . . : actually it should have been of Science, Technology and Medicine, except the name would have been too cumbersome. Or, perhaps it could have been called simply: A Colloquium devoted to the Historical Sociology of Scientific Knowledge.

Like the parent institution, situated in the intellectual Mecca of Boston, the Israel Colloquium is at the crossroads of various kinds of contradictory attitudes, polarized worldviews and a multitude of intellectual trends which brings scholars from many nations – alas not yet from all – to visit Israel and to conduct a dialogue on their topic of interest.

In the academic year 85/86 the Israel Colloquium has started its fifth annual lecture series, and it has been gathering momentum and recognition with the scientific community. The first contract for publication of its selected proceedings was signed with Dr. Simon Silverman, the founder and moving spirit of the Humanities Press, a man who did so very much since 1945 to promote our field in the United States and recently also in Israel with publications and direct donations. After his sudden death in 1984, which we mourn deeply, the Humanities Press underwent changes and the Israel Colloquium volumes appropriately found a home with the D. Reidel Publishing Co., among the *Boston Studies in the Philosophy of Science* where books from the Boston Colloquium have appeared from its beginning.

We hope that this new series of scholarly books will prove a worthwhile contribution to the study of Science as a Cultural System.

September 1985

ROBERT S. COHEN
Boston University
YEHUDA ELKANA
Tel-Aviv University
MARX W. WARTOFSKY
Baruch College, City University of New York

Contents

PREFACE

This collection is the first proceedings volume of the lectures delivered within the framework of the Israel Colloquium for the History, Philosophy and Sociology of Science, in its year of inauguration 1981-82. It thus marks the beginning of a new venture.

Rather than attempting to express an ideology of the unity of science, this collection in fact aims at presenting a kaleidoscopic picture of the variety of views about science and within science. Three main disciplines come together in this volume. The first of scientists, the second of historians and sociologists of science, the third of philosophers interested in science. The scientists try to present the scientific body of knowledge in areas where the scientific adventure kindles the imagination of the culture of our time. At the same time, of course, they register their own reflections on the nature of this body of knowledge and on its likely course of future development. For the historians and sociologists, in contrast, science is there to be studied diachronically, as a process, on the one hand, and synchronically, as a social institution, on the other. As for the philosophers, finally, their contribution to this series is not meant to remain within the confines of what is usually seen as the philosophy of science proper, or to be limited to the analysis of the scientific mode of reasoning and thinking: it is allowed, indeed encouraged, to encompass alternative, and on occasion even competing, modes of thought.

The collection of essays in this volume, then, does not present one central sustained argument. Rather than *argue* a case, these essays, taken together, *exemplify* it, through the juxtaposition of a variety of points of view of science both as a system of propositions and as a social institution. The metaphor of the kaleidoscope is thus peculiarly apt. As an alternative metaphor one might think of a cubist painting that attempts to represent a plurality of viewpoints of a certain object on a single plane.

And, if an educational aim is sought for this volume, it could perhaps be expressed in Tennyson's words —

> Nourishing a youth sublime
> With the fairy tales of science, and long result of Time.

As is well known to those experienced in such matters, the running of an annual series of lectures and the production of a proceedings volume are enterprises which exact — and depend on — the good will and cooperation of many people. The largest and deepest debt of gratitude is due to Yehuda Elkana, the prime mover of the entire Israel Colloquium venture. Thanks for indispensable editorial, administrative and secretarial help go to Robert Amoils, Yael Avner, Mordechai Dwek, Edna Gil, Rachel Kerem, G. H. Schalit, Hanna Shapira, Esther Shashar, and Fred Simons.

Edna Ullmann-Margalit   Jerusalem
The Coordinator

On the Empirical Application of Mathematics and Some of its Philosophical Aspects

STEPHAN KÖRNER

Among the distinctive features of mathematical structures or theories is their actual or potential applicability to empirical phenomena. It is the purpose of this essay to compare mathematical structures, especially those of arithmetic and the real number system, on the one hand, with empirical structures, especially those of discrete and continuous phenomena, on the other, to examine what is involved in applying mathematical to empirical structures, and to exhibit some metaphysical assumptions about their relations to each other. The essay begins with some remarks on what might be called the "empirical arithmetic of countable aggregates" (Section 1). Next an indication is given, how this empirical arithmetic is idealized into a pure arithmetic of natural numbers and integers, and, beyond, into a pure mathematical theory of rational and real numbers (Section 2). There follows a brief characterization of empirical continua (Section 3); a discussion of the application of pure numerical mathematics to empirically discrete and empirically continuous phenomena; and some remarks on the application of mathematics in general (Section 4). The paper ends by drawing attention to some relations between pure and applied mathematics on the one hand and metaphysics on the other (Section 5).

E. Ullmann-Margalit (ed.), The Kaleidoscope of Science, 1–11.
© *1986 by D. Reidel Publishing Company.*

1. On the Empirical Arithmetic of Countable Aggregates

Whether or not Frege's analysis of the number-concept is sufficient as a foundation of (pure) arithmetic — an assumption which he himself rejected toward the end of his life — it clearly draws attention to what is involved in counting empirical objects. In order to do so, one must be able to identify the objects, which are to be counted, as particulars; to collect them into aggregates; to determine the equinumerousness or otherwise of different aggregates; and to produce or reproduce a fundamental sequence by reference to the subsequences of which the number of any empirical aggregate can be determined (as equinumerous with a particular subsequence). The mentioned empirical concepts of the identification and aggregation of particulars, of equinumerousness and of a fundamental sequence lack the precision and exactness of the corresponding Fregean notions.

Thus the identification of an empirical object need not be achieved by determining its membership in a class and by distinguishing it from the other members, if any, of this class by an attribute which it does not share with them. For an empirical particular may be successfully identified by being incorrectly classified, e.g. when the dog in the corner, which is to be counted as belonging to a given aggregate, is misclassified as a cat. Again, when collecting particulars into an aggregate for the purpose of counting, it is not necessary that the aggregate be a class, whose members are counted, a complex particular, whose parts are counted, or ambiguously either a class or a complex particular.

The determination of the equinumerousness or otherwise of two aggregates, both of which are perceptually present, is particularly easy. The same applies to counting the elements of a perceptually present aggregate by comparing it with a subsequence of a perceptually present fundamental sequence, e.g. an aggregate of horses, which turn out to be three in number, when matched with the corresponding subsequence of the fundamental sequence consisting of, say, the fingers on the two hands of a primitive tribesman, taken in an agreed succession. If this finite sequence is not sufficient for the counting of empirical objects, it can be either prolonged or stopped by calling the number of any aggregate which is larger than the fundamental sequence "larger than the last number". "many" or by another name which indicates an inability to count beyond a certain number, a lack of interest in doing so, or both. Here one may without loss ignore this possibility and assume that a fundamental sequence excluding it always is — or can be made — available.

A fundamental empirical sequence is in an obvious manner ordered by the relation of succession: It contains a member which has no predecessor. It also contains a member which has no successor. Every other member has a unique predecessor and a unique successor. On the basis of this observation one can define the empirical number *one* as the number of the subsequence containing the predecessorless member and of any aggregate equinumerous with this subsequence; the empirical number *two* as the number of the subsequence containing the predecessorless member and its successor, and of any aggregate equinumerous with this subsequence, etc., until one arrives at the successorless subsequence. The successor relation within the subsequences manifests itself also in their numbers, so that if two empirical numbers are the same, their predecessors and their successors (if any) are also the same. There is no need to show how 'addition', the relations 'less than' and 'multiplication' can be defined in our empirical arithmetic. Nor is it necessary to show that the resulting arithmetic will be affected with certain imprecisions and that because of the inexactness of the empirical concepts used in developing it, the number-concepts admit borderline cases.

2. On Pure Arithmetic

The most important step by which pure arithmetic achieves the exactness of its concepts and operations as well as the ability to overcome the limitations based on the finiteness of all empirical aggregates, is the replacement of an empirical by an ideal fundamental sequence. Of the various possible ones the sequence of von Neumann is particularly well suited for our purpose, since it can be constructed transparently in a succession of steps such that if a step is taken, the next step may, but need not, be taken. The construction is based on the concept of an exact class, including that of the empty class (e.g. as defined by the requirement that it contains all square circles), and the concept of one class including another.

In order to take the first step in constructing the von Neumann fundamental sequence, one postulates: (i) that the sequence has one and only one member which has no predecessor and at most one member which has no successor; (ii) that the empty class, briefly Λ, is the predecessorless member; and (iii) that the successor of a member of the sequence is the class containing the preceding members as its elements. So far the sequence thus has the following form:

$$\Lambda, \{\Lambda\}, \{\Lambda\{\Lambda\}\}, \{\Lambda\{\Lambda\}\{\Lambda\{\}\}\} \ldots \tag{1}$$

where the dots indicate the continuation of the sequence up to the successorless member, if its existence is postulated, or the endless continuation, if its nonexistence is postulated. In the former case our fundamental sequence and the arithmetic based on it are finitist. Clearly a finite von Neumann sequence may be "larger" than any empirical aggregate (in a sense of the term which presupposes the applicability of pure finite arithmetic to empirical aggregates).

Once in possession of the arbitrarily large von Neumann sequence, one may define the corresponding natural numbers

$$0, 1, 2, 3, \ldots \tag{2}$$

as respectively equinumerous with the members of the sequence in the order of their occurrence in it. A possible further step which is taken by almost all mathematicians is to postulate (iv) that no member of the fundamental sequence and, thus, no natural number be without a successor, i.e. that the sequence of natural numbers be in some sense "endless" or "infinitely proceeding." In order to indicate the difference between the structure of finite or arbitrarily large sequences on the one hand and of endless progressions on the other, one might express the incompleteness of the former by *etc.* and the incompleteness of the latter by *ETC.* The question of the manner in which the sequence

$$0, 1, 2, 3, ETC. \tag{3}$$

is given, is either not raised at all or (rightly) considered philosophical rather than mathematical. The so-called constructivists admit the infinitely proceeding sequence of natural numbers and other infinitely proceeding sequences into mathematics, while rejecting the assumption that the members of such sequences can in any sense be given in their totality. They consequently have to reject the infinite totalities which are used by most mathematicians in the arithmetic of natural numbers, integers, fractions and — most important — real numbers.

The constructivists thus do not take the further idealizing step, which is taken by the majority of mathematicians in their conception of natural numbers either independently or via their conception of a fundamental sequence. This is to postulate (v) that there is available a set

$$\{0, 1, 2, 3, ETC, \}, \tag{4}$$

i.e., to postulate "as given a set, i.e., a totality of things called natural numbers."[1] In accordance with a widely accepted terminology one might call the sequence

$$0, 1, 2, 3, ETC. \tag{3}$$

which satisfies the first four postulates, the constructivist sequence of natural numbers; and the totality

$$\{0, 1, 2, 3, ETC.\} \tag{4}$$

which in addition satisfies the fifth postulate, the classical, the non-constructivist or Platonist totality of natural numbers. The constructivist sequence serves as a foundation of constructivist arithmetic, and the classical totality as a foundation of classical arithmetic.

Just as empirical arithmetic, involving the empirical relation 'less than' and the empirical operations of addition and multiplication, is founded upon the availability of a fundamental, empirical sequence, so constructivist and classical arithmetic are, respectively, founded on the fundamental constructivist sequence or on the totality of natural numbers. And since both of them are idealizations of empirical fundamental sequences, both of them are idealizations of empirical arithmetic. Classical arithmetic, though a more radical idealization of empirical arithmetic than constructivist arithmetic, is nevertheless more familiar than the former.

The axioms of the classical arithmetic of natural numbers are found in textbooks on the foundations of mathematics and include axioms governing addition, multiplication, the ordering relation 'less than,' as well as an axiom connecting addition and multiplication, namely $x(y + z) = xy + xz$, axioms connecting addition and ordering, e.g. $(x < y)$ *implies* $x + z < y + z$, and axioms connecting multiplications and ordering, e.g. *if* $z > 0$ *then* $x < y$ *implies that* $xz < yz$. In pure arithmetic of natural numbers not every equation of form $x + n = m$ and not every equation of form $x \cdot n = m$ have a solution. In order to close the first gap one must extend the totality of natural numbers into the totality of (positive and negative) integers and show that the properties which characterize the addition, multiplication and ordering of the natural numbers also characterize the integers. While the series of integers

$$\{\ldots, -3, -2, -1, 0, 1, 2, 3, \ldots\} \tag{5}$$

obviously differs in structure from the sequence of natural numbers by having no first member, the two series are equinumerous which can be seen, e.g., by matching the odd natural numbers with the positive integers and the even natural numbers with the negative integers. Calling the cardinal number of the infinite totality of all natural numbers and of every equinumerous set $\aleph_0$ (aleph-zero), one may also say that the set of all integers is of the same size or has the same cardinal number as the set of all natural numbers. Just as constructivists reject any infinite totality, so some mathematicians reject any infinity greater than $\aleph_0$. [See e.g. Hermann Weyl, *Das Kontinuum* (Berlin, 1918).]

There is no need here to show in detail how the classical theory of natural numbers is extended into the classical theory of rational and real numbers or to compare the thus extended theory with its intuitionist counterpart and other nonclassical theories. For our purpose it is sufficient to recall that in the classical theory of rational numbers the set of rational numbers between 0 and 1 has the cardinal number $\aleph_0$ and is dense; and that the set of real numbers between 0 and 1 has the cardinal number $2^{\aleph_0}$ and is continuous. Descartes' program of reducing geometry to arithmetic and Leibniz's even more radical program of reducing all necessary propositions and, hence, both arithmetic and geometry to logic, require the characterization of all continua, including perceptual ones, as *mathematically* dense and continuous, i.e. as consisting not only of an infinite totality of elements but of a nondenumerably infinite totality of elements.

3. On Empirical Continua

Empirical or perceptual continua are not infinitely divisible but only finitely divisible. Moreover, what characterizes such a continuum — be it a drawn line, a time interval, a painted color spectrum, a continuous movement — is that it can be divided into two gapless subcontinua in such a manner that their common border exists only *qua* border of the subcontinua, and that this process can be continued a finite number of times depending on the percipient's (finite) powers. The distinction between two kinds of parts of a continuum, namely subcontinua and common borders of subcontinua, was emphasized by Aristotle and, in a more precise way, by Brentano.[2] Yet, even Brentano's notion of a common border of two subcontinua which, unlike these, has no independent existence, stands in need of further analysis.

Such an analysis can be given if one abandons Frege's principle of exactness which requires that "the definition of a concept ... determine for every object unambiguously ... whether or not it falls under the concept."[3] The principle admits only the following two relations between a concept and an object: (1) It is correct to apply the concept to the object and incorrect to refuse the concept to the object; (2) It is correct to refuse the concept to the object and incorrect to apply the concept to the object. It excludes two further possibilities, namely: (3) It is correct to apply the concept to the object and correct to refuse the concept to the object (but not to do both); (4) It is incorrect to apply the concept to the object and incorrect to refuse the concept to the object. If the third possibility is realized, the object is a neutral or borderline case of the concept. The fourth possibility is that of the object's being an undecidable case.

One can develop a logic in which, apart from exact predicates which have only positive and negative instances, there occur also inexact predicates, which have positive, negative and neutral instances. The occurrence of such predicates is acknowledged by Frege even though he regards them as expressing "concept-like formations" which "logic cannot recognize as concepts" for which "exact laws can be stated" (Frege, *loc. cit.*). Two inexact predicates may share all or some of their neutral cases. Thus 'x is green' and 'x is not green' share all their neutral cases, whereas 'x is green and heavy' and 'x is green' share only some of them.

By combining inexact predicates one can form new predicates, including "degenerate" ones which have no positive, no negative or no neutral instances. Among them the border-predicates, as one might call them, are here of special interest. A predicate $B(F, G)$ is a border-predicate for F and G (characterizes the common border between F and G) if, and only if, its positive instances are the common neutral instances of F and G and its negative instances all other objects. A border-predicate — e.g. *B (green, not green), B (x lies on the left side of a certain dividing line of a perceived line segment, x lies on the right side of this dividing line)* — has no neutral cases. It does not characterize a subcontinuum, but the common border which connects two subcontinua, and it is defined only as their common border. Brentano's characterization of continua as containing two kinds of parts, namely subcontinua, e.g. parts which can exist independently of the continuum of which they are parts, and parts which exist only as borders, is thus made more precise by distinguishing between inexact predicates applying to the former and (exact) border-predicates applying to the latter. The latter distinction does not imply that a continuum has an infinite number of parts or be divisible *ad infinitum*.[4]

A more elaborate analysis of empirical continua might, among other things, involve a distinction between continua of one or more dimensions. Thus one might define an empirical continuum as one-dimensional if, and only if, no border between any of its subcontinua is itself a continuum; an empirical continuum as two-dimensional if, and only if, at least one border between two of its subcontinua is one-dimensional and every other border is either one-dimensional or no continuum; and as n-dimensional if, and only if, at least one border between two of its subcontinua is $(n-1)$-dimensional, and every other border is less than $(n-1)$-dimensional or no continuum. Examples of one-, two- and three-dimensional continua are, respectively, empirical line-segments, planes and bodies. Continua of higher dimensionality might be produced by combining spatially continuous phenomena with such as exhibit continuous gradations of color, weight, motion, etc. And one might point to other parallels between mathematical and empirical continua.

But, it may be asked, are not the continuous structures which can be formally characterized by means of a finite logic of exact and inexact predicates themselves ideal rather than perceptual structures? I prefer to leave this question open. Yet, whatever the answer, it will be sufficient to agree that mathematically dense or continuous structures are radically different from — and (in a sense to be made clearer) idealizations of — empirical continua; and that the structures characterized in terms of a finite number of subcontinua and borders or a finite number of exact and inexact border-predicates at the very least preserve, and draw attention to, *some* features of empirical continua which are no longer present in mathematically dense or continuous sets.

4. On the Application of Pure Numerical Mathematics to Empirically Discrete and Continuous Phenomena and the Application of Mathematics in General

The application of the pure arithmetic of natural numbers and integers to empirically given particulars and aggregates is a species of idealizing representation. It consists mainly in identifying an ideal fundamental sequence with an empirical one, even though the former may be larger than any finite, empirical sequence, may be infinitely proceeding or may be infinitely proceeding and "given" as a complete totality; and in similarly identifying the empirical less-than relation, sum, and product with the corresponding ideal relations and operations. This identification is, like any other representation, justified by its purpose and context. The application of the pure mathematics of rational and irrational numbers to empirical continua consists in identifying even more disparate entities, since empirical continua are not sets of distinct elements whose cardinal number is $\aleph_0$ or even $2^{\aleph_0}$.

The preceding account of the application of pure arithmetic and the mathematical theories of rational and real numbers, can be extended to other mathematical theories and their application to empirical phenomena. It fits in well with the manner in which empirical inquiries may inspire the creation of new mathematical theories. For these inquiries frequently involve the search for a representation of empirical phenomena such that on the one hand the *representans* is simpler or in other respects structurally different from the *representandum*, and such that on the other hand the structural differences between *representans* and *representandum* are for the purpose of the representation wholly irrelevant or, at least, negligible.

It is worth noting that the dominant, modern philosophies of mathematics give altogether different accounts of the application of mathematics to experience. According to Frege (in his logicist phase) and his logicist successors, mathematics is logic and is thus, like logic, implicit in all consistent thinking so that the problem of the application of mathematics is no more and no less a problem than that of the application of logic. According to Kant, and also Brouwer and his intuitionist school, pure mathematics *describes* perceptual structures and constructibilities, there being thus no difference between pure and applied mathematics. Similarly, according to Hilbert and his school, finite mathematics is the description of some aspects of perception while infinite mathematics is strictly meaningless. None of these views thus takes any notice of the gap between the description and mathematical idealization of empirical structures and the manner the gap is bridged in the application of mathematics.[5]

The acknowledgment of the gap between mathematical and perceptual structures and the thesis of its being bridged by as-if identification recall two rather different philosophical traditions. The first is that of Plato who distinguishes between perception, which "tumbles about between being and not-being" and Reality, which includes the Forms of mathematics. He conceives of the relation between the two as μέθεξις (participation, approximation, etc.), which resembles as-if identification. Plato, of course, holds that there is only one Reality, only one Platonic heaven, whereas the view propounded in this paper implies that there are many such. The second tradition is that of Leibniz, who regards the notions of the infinitely large, the infinitely small and some other mathematical notions as mathematical fictions or useful idealizations, which can be eliminated after being used. He does, of course, conceive of the core of mathematics as nonfictitious logic.

5. On Some Metaphysical Constraints on the Content, the Logical Form and the Ontological Status of Mathematical Theories

The preceding remarks imply that mathematical thinking involves

metaphysical principles, i.e. a species of transdisciplinary principles which
are supreme in the sense that their acceptor requires that all his beliefs be
consistent with them.[6] A person's metaphysics may thus affect his
mathematics. For example, if he considers two axiomatized mathematical
theories, the theorems (and axioms) of which overlap, and if he notices that
one of these theories is inconsistent with his metaphysical principles, then he
will reject it on metaphysical grounds. A well-known example is the rejection
by Kantians of non-Euclidean geometry as metaphysically incorrect,
because it is inconsistent with the nature of space as determined by their
metaphysical principles. In this connection it does not matter whether they
regard the parallel postulate (or one of its deductive equivalents) as
belonging to metaphysics, geometry or both. Similar remarks apply to
alternative mathematical theories of natural numbers, integers, and rational
and irrational numbers, to which a person's metaphysics is relevant insofar
as it includes principles about the nature of a fundamental sequence as finite,
as infinitely proceeding, as infinitely proceeding and completely given, as
being completely given with or without its power-sets also being completely
given, etc.

A second metaphysical issue relating to mathematics concerns the logical
status of the ideal structures which are its subject matter. That is to say, one
may ask and in different ways answer the metaphysical question, whether
mathematics is about particulars; about attributes of particulars; or
indifferently about either. Thus Plato held that all mathematics is about
particulars, namely the Ideas or Forms; Kant that (Euclidean) geometry is
about a particular entity, namely the pure intuition of space, and that
arithmetic (as based on a merely potentially infinite, fundamental sequence)
is about another such entity, namely the pure intuition of time; and Leibniz
held that arithmetic and geometry are about attributes, namely temporal and
spatial relations between particulars. Modern set-theory conceives
mathematical structures as sets of individuals standing in certain relations. It
is nevertheless compatible with both positions, since the individuals can be
conceived both as real individuals or as dummies whose only funcion is to
characterize relations.

The applicability of at least some mathematical theories, i.e. the
identifiability of empirical with mathematical structures in certain contexts
and for certain purposes, raises a third metaphysical issue relating to
mathematics. It concerns its ontological status, e.g. the answer to the
question whether mathematical structures exist independently of empirical
structures or whether they exist only, or mainly, as their idealizations, i.e.
only or mainly insofar as they are identifiable with the empirical structures of

which they are idealizations. The answers to these questions cover a wide spectrum. At one extreme lies the radical empiricist answer (of e.g. Berkeley) that insofar as mathematical structures differ from perceptual ones, they are at best fictions. At the other lies the radical rationalist answer (of e.g. Plato) that only mathematical structures are genuinely "real" and that empirical phenomena can only be conceived as imperfect approximations to them. Between these extremes lie views which regard some mathematical structures as independently existing particulars or as characterizing such particulars and others as fictions. An example is Hilbert's formalism according to which finite arithmetic and combinatorics are true of the empirical world (or a negligibly idealized version of it), while infinitistic mathematics is fictitious.

In conclusion, it seems appropriate briefly to recall the following main theses of this essay. First, the problem of the application of mathematical to perceptual structures arises from the fact that — contrary to the teachings of logicists, formalists and intuitionists — the structures are not isomorphic. Second, the application of mathematics to perception consists in treating an applied mathematical structure and an empirical structure to which it is applied (for certain purposes and in certain contexts) *as if* they were identical. The principle on which the identification is based might, in contrast with Leibniz's principle of the identity of indiscernibles, be called "the principle of the identifiability of discernibles." Third, insofar as mathematical structures are idealizations of empirical structures, there is room for a variety of mutually inconsistent mathematical structures which are yet applicable to the same empirical structure.

Notes

1   The quoted words form part of the first sentence of the first chapter of a book by E. Landau explaining the calculation with whole, rational, irrational and complex numbers. See *Grundlagen der Analysis* (Leipzig, 1930).
2   See Aristotle, *Physics*, book VI; and Brentano, *Raum, Zeit und Continuum* (Hamburg, 1976).
3   See *Grundgesetze der Arithmetik*, vol. 2, p. 69 (Jena, 1903).
4   For a more detailed discussion of empirical continuity, see chapter IV of my *Experience and Theory* (London, 1966); and for a formal analysis, see J.P. Cleave, "Quasi-Boolean Algebras, Empirical Continuity and Three-Valued Logic," *Zeitschr. für Math. Logik und Grundlagen d. Mathematik* 22 (1976):481–500.
5   From Hilbert's formalism one should distinguish the formalism of A. Robinson, whose view on the application of mathematics is similar to the one which has been taken here. See his "Formalism 64" and "Concerning Progress in the Philosophy of Mathematics," in: *Selected Papers*, vol. 2 (New Haven, Conn.: Yale University Press, 1979).
6   For a more detailed account, see e.g. "Science and the Organization of Belief," in: *Science, Belief and Behaviour*, ed. D.H. Mellor (Cambridge, Mass.: Cambridge University Press, 1980), pp. 43–61.

On the Empirical Application of Mathematics
A Comment

HAIM GAIFMAN

I

Professor Körner claims to have discovered a gap between pure mathematics and empirical structures that has been hitherto overlooked by all the major schools in the philosophy of mathematics. This discovery is summed up in his statement that "contrary to the teachings of logicists, formalists and intuitionists, the structures are not isomorphic." Empirical structures, so the argument runs, involve irreducible vagueness, ambiguities and border cases, and they do not necessarily constitute precisely defined classes. An empirically ordered aggregate may serve to represent a finite fragment of natural numbers in the usual way: Associate the number one with the subsequence consisting of the first member and its successor and so on. We are told that the basic arithmetical operations can be defined in such an empirical model but that, owing to the inexactness of the empirical in such an empirical model but that, owing to the inexactness of the empirical concepts, the resulting arithmetic "will be affected with certain imprecisions." One is curious to see samples of this imprecise arithmetic. Perhaps the author had in mind the representation of different numbers by the "same" sequence, owing to its imprecise definition; say, the sequence of apples that I am seeing now may represent either three or two, depending on

E. Ullmann-Margalit (ed.), The Kaleidoscope of Science, 13–16.
© 1986 by D. Reidel Publishing Company.

whether that pear-like object is considered to be an apple or not. One wonders where to draw the line that separates legitimate ambiguous statements of "empirical arithmetic" from simple errors. Under what conditions should both assignments of the two numbers to our sequence be legitimate? Should the doubtful object be a real border case of apple–non-apple, undecidable by scientific analysis? Or should it only be reasonably misleading? May counting mistakes, such as the obtaining of different results in two countings of the same sequence, produce legitimate ambiguities of "empirical arithmetic"? How does all this affect the correctness or incorrectness of statements such as $2+3=5$, $2+3=4$? It becomes clear that there is no one "empirical arithmetic" but that every particular empirical structure, on each particular occasion, may give rise to a sort of "empirical arithmetic" (whatever that may be).

On the other hand, pure mathematics represents all natural numbers by means of one ideal fundamental sequence, thus achieving the precision lacking in the empirical model. The underlying structure is not isomorphic to empirical ones, hence the gap. This gap can be bridged because in many contexts the differences between the structures are of no consequence. We can therefore treat the empirical structure *as if* it were the ideal one.

Thus far — what I take to be Körner's main contention.

Let me remark at the outset that arithmetic does not depend for its meaning or validity on the possibility of pinpointing one particular fundamental sequence. Such a sequence may serve as a convenient tool, but to represent it as "the most important step by which pure arithmetic achieves the exactness of its concepts and operations" is misleading. Arithmetical concepts had been clearly and correctly understood hundreds of years before von Neuman's definition of his fundamental sequence (or the other variants proposed before it). But this is a minor point.

Now, it is trivially true that abstract mathematical entities are different from empirical aggregates. But by no means does this imply that the empirical application of arithmetics rests on *as-if* identifications. Körner is, I think, confusing two sorts of similarities: (i) A circle, *C*, drawn on a sheet of paper, can be exact enough so as to allow the application of abstract geometrical rules in various measurements relating to it. These rules are valid for ideal circles, which *C* is not. But it resembles one to the extent of enabling us to treat it *as if* it were ideal. Here is a case of *as-if* identification. (ii) The aggregate of strokes ||||||||||, the aggregate of circles ∘°∘ ∘°∘ ∘°∘, the set {0, 1, 2, 3, 4, 5, 6, 7, 8} and the set of planets bear no resemblance to each other, but they share one common feature. All contain the same number of objects — the number 9. This feature is *not approximated to some degree*; in all cases it is *exactly the same*. However, according to Professor Körner, in asserting that the first aggregate contains 9 strokes, we have idealized it. The real "empirical number," whatever it may be, is different; it resembles its ideal counterpart sufficiently, just as the circle C resembles an ideal circle. This, I claim, is simply false. In fact, the "ideal" 9 is the very feature that is shared by the various aggregates, empirical or not. We *do* have a precise isomorphism — in this case it is a one-to-one mapping between equinumerous sets.

There are of course borderline cases, say the group of all people in some room, with someone standing in the doorway. But their existence does not invalidate the clear-cut standard examples.

To claim at this point that vagueness, because it is inherent in empirical contexts, cannot be excluded in principle, is to put the cart before the horse. It is only after the empirical data have been organized that vagueness and borderline cases make any sense. And number-theoretic concepts are an inevitable constituent of this organization. Elementary perceptions, such as those expressed by the statement "here are three coins," precede and must be presupposed by any later awareness of borderline cases. The "three" indicates here the precise mathematical three. It enters as an irreducible constituent, not as an idealized "empirical three."

This analysis, essentially Kantian, is certainly not new. Neither is the interpretation of very elementary statements (such as: "Here are 10 apples" or "There are more pupils in class A than in class B") an issue of controversy among the major schools in the philosophy of mathematics. Frege's analysis of such statements, in his *Grundlagen der Arithmetik*, stands. I do not think that it has been invalidated or that a better one has been offered since. (The contradiction in his formal "Grundgesetze" system is of course quite a different matter.)

Where the major philosophies of mathematics begin to differ is in the interpretation of the infinite totality of numbers, in other words — the meaning of statements relating to all natural numbers, such as "there are infinitely many primes."

II

Professor Körner uses "empirical" in a very elementary sense, i.e. to indicate basic perceptions, such as those expressed by statements of the kind discussed above: "There are three apples in the basket," etc. Even here his "empirical arithmetic" is out of place. It is even more so when confronted with everyday statements such as: "The inflation rate has jumped to 8.2%

last month," or "This computer has 300,000 directly accessible memory locations, each containing 16 binary bits," or "There are more than 10 billion possible combinations of the Hungarian cube." These are a few of the numerous examples where a highly abstract arithmetical machinery enters directly into empirical everyday contexts. Is there any "empirical arithmetic" that is being idealized in, say, the last statement concerning the Hungarian cube? If so, what is it? Shall we visualize an empirical aggregate of 10 billion cubes each realizing a different combination? Or shall we imagine a temporal sequence in which one cube is being run through all its combinations? Assuming that we have a clear notion of "combination" (and this is not difficult), the statement that there are 10 billion combinations is clearer and more fundamental than any proposed explication in which this number is "empirically realized."

Note that very simple toys can be invented that give rise to combination numbers exceeding the estimated number of atoms in the universe or the number of minutes in the galaxy's life span. It is hopeless to attempt to give a meaning to "the number of combinations of toy X" through some empirical realization of it. And yet a 12-year-old child can understand it clearly.

III

Professor Körner suggests a different kind of logic, along the lines proposed by him in his book *Experience and Theory* and developed technically in Cleave's paper "Quasi-Boolean Algebras, Empirical Continuity and Three-Valued Logic." This logic is to reflect the possibility of border cases insofar as $P(a) \wedge \neg P(a)$ is not considered contradictory. (On the other hand, equalities $x = y$ are treated in the standard way, i.e. $x = y \wedge \neg (x = y)$ is contradictory.) Without going into the details, let me state that this logic does not lead to any new kind of arithmetic. In geometry it amounts to a reformulation of the standard system in a way that reflects better the intuitions expressed in the passages quoted from Aristotle and Brentano. But the system is essentially the standard one in a different clothing, as is established by easy translations in both directions.

The above comments are not intended as a criticism of Cleave's paper which is technically competent and may serve as a nice explication of some intuitions concerning the concepts of part and boundary. I only wish to point out that it has no philosophical import either for the foundations of arithmetic or for geometry.

Meaning and Our Mental Life

Hilary Putnam

The thrust of this lecture will be negative: I shall argue that a certain way of thinking about meaning and about the nature of the mind is fundamentally misguided. It is always less exciting to hear someone criticize attempted solutions to a problem than to hear someone announce that he has found the solution. But I think we can learn something about the nature of meaning and, perhaps, something — even if it is somewhat nihilistic — about the nature of psychology by seeing why certain ideas about meaning and its place in the mind don't work.

The way of thinking I am going to criticize comes from ideas of Noam Chomsky as interpreted and extended by Jerry Fodor in *Language and Thought*.[1] I do not know to what extent Chomsky actually agrees with Fodor's theories; sometimes it seems as if he more or less agrees with Fodor, and at other times it seems as if Chomsky has really distanced himself a bit from the view I am going to criticize. However, there is a widespread expectation that Chomsky's ideas will sooner or later *be* extended to the realm of semantics, an expectation which is responsible for the use that is made of his ideas by French neo-structuralists, American cognitive psychologists and others, and this may justify my focusing on Fodor's book as representative of this expectation. Chomsky refers at times to a level of "semantic representations" in the mind, which suggests that he himself shares the expectation in question.

E. Ullmann-Margalit (ed.), The Kaleidoscope of Science, 17–32.
© *1986 by D. Reidel Publishing Company.*

Chomsky is famous for having proposed a theory according to which grammar is "innate" in the mind. According to this theory, there is a universal grammar — a structure and a set of categories which are universal, not just contingently, not just because human environments are in certain respects alike, but innately: a syntax built into the basic structure of the mind itself.[2] Recently, Chomsky has suggested that this innate linguistic structure characterizes not the whole mind but the way of functioning of a particular "module" in the mind, the so-called "language organ."[3] Chomsky and Fodor appear to conceive of the language organ as a relatively "dumb" organ, independent of general intelligence (if there is such a thing). (This seems to be a sharp turn from Chomsky's earlier papers and books, which appeared to stress the model of the mind as learning its native language — with the aid, indeed, of its innate knowledge of "universal grammar" — via something like *hypothesis formation.* The more recent picture of the mind as a collection of automatically functioning "modules" with its stress on "bottom up" as opposed to "top down" processing — that is, on automatic processing as opposed to processing which draws on general intelligence and general information — seems surprisingly similar to some of the tendencies in the Behaviorism that Chomsky so sharply attacks.)

At any rate, given that the key ideas of Chomsky's theorizing are the ideas of linguistic universals, of innateness, and (more recently) of modularity, the form that one can expect a Chomskian theory of the semantic level to take is relatively clear (and Fodor's theory does take this form), even if the details can take various different shapes. A Chomskian theory of the semantic level will say that there are "semantic representations" in the mind/brain (in Fodor's theory these are formulas in a hypothetical language of thought, often referred to as "mentalese"); that these are innate and universal; and that all our concepts are decomposable into such formulas or "semantic representations." This is the theory that I hope to destroy.

Some of you know that I am also sceptical about the idea of innateness as applied to syntax,[4] but I am not going to say anything about that today. Chomsky's work, and especially Chomsky's revival of "mentalism" and his talk of universals in language, has excited worldwide attention, and I do not think this is just because people are interested in syntax. As I have already mentioned, these ideas have attracted attention not only in the English-speaking world but also in the French-speaking world, and the attention of people very remote from any concern with technical linguistics: Lacanian psychoanalysts, anthropologists, philosophers of all kinds. Obviously, people do anticipate that the Chomskian approach should pay off with respect to something other than grammar.

I would not try to destroy the theory of innate "semantic representations," but would rather content myself with simply ignoring it if I did not think that there is much to be learned from studying the questions that it raises and the answers it proposes, and if I did not also think that the brilliant thinkers who propound such theories are in the grip of an intellectual desire which is itself a phenomenon worth taking seriously. The desire is one that springs, I think, from two tendencies in the history of recent thinking about the mind.

One tendency is a tendency, or a yearning, to rehabilitate the oldest pattern of explanation of our mental workings there is: explanation in terms of beliefs and desires, or, as philosophers say, in terms of "propositional attitudes." No matter how strongly the tides of Behaviorism may have run, we have never been able to stop explaining our own behavior and that of others in terms of propositional attitudes. We say: "I went to school today because I had to teach a class," or "I went to the market because I knew we were out of milk, and I wanted to be sure we had milk for breakfast."

These explanations can easily be cast into the form that Aristotle called "the practical syllogism," that is, the standard pattern of belief-desire explanation. Behaviorism in its most radical form suggested that we don't need any of that, because all that we are really talking about is conditioned responses, habits, stimuli, etc. Perhaps one can replace belief-desire talk by stimulus-response talk when one is dealing with rats in very controlled situations, but even Skinner ran into trouble when he tried to use stimulus-response concepts to describe daily life (including verbal behavior) outside of such controlled situations. What Skinner had to do was, basically, to widen the notions of response and stimulus so that (as Chomsky pointed out in a famous review some years ago[5]) they became empty. Skinner, in the course of trying to analyze an utterance about World War II, referred to the whole war as a "stimulus." Chomsky pointed out, quite reasonably, that once the notion of a "stimulus" becomes so wide that World War II can be a stimulus (and the response can take place 40 years later), stimulus-response talk has become a mere jargon with no real control. So there are certainly good reasons to go back and rehabilitate belief-desire explanation.

The other tendency is the tendency to think of the brain as a computer and of our psychological states as the software aspect of the computer. In such views (e.g., in "artificial intelligence" work), it is generally assumed that the brain has a built in (and thus "innate") formalized language which it can use both as a medium of representation and as a medium of computation. (The idea of a *lingua mentis,* a language of the mind, is really an old idea that has made a reappearance, somewhat like the idea of a Beginning of the Universe.)

If we identify the computer's *lingua mentis* with the level of "semantic representations," we arrive at the following picture: when we learn the meanings of the words in our first language, for example, when the French child learns that the word *mortel* can mean "mortal" or "deadly dull," or when the American child learns the meaning of the word *carburetor*, or when the Israeli child learns the meaning of the word *shemen* ("oil"), what is learned in each case is something like a definition in or a translation into "mentalese." This picture revives what used to be called the Post Office model of the mind. The mind thinks its thoughts in "mentalese," codes them in the local natural language, and then "transmits" them to the Post Office of the other speaker, who thereupon proceeds to decode the "letter," and extract the thoughts. In this picture, language, in the sense of natural language, far from being the essence of thought, is a mere vehicle for the communication or expression of thought.

These ideas, the idea of reviving mentalism and the idea of a computational model of the mind, can appeal for many reasons. For example, the idea of a structure common to all human minds has eighteenth-century reverberations. If Chomsky is right, all mankind has a single nature, apart from deformations. Chomsky connects this with the Enlightenment, and with the political ideals of liberty, equality, and fraternity.[6] But even apart from such reverberations, it is natural that many thinkers should feel attracted to a program which brings together mentalistic psychology and computational psychology. The desire to bring these two together is one that gripped me, too, for a long time. These two tendencies are the dominant anti-Behaviorist tendencies, and one might plausibly think that they would gain enormous strength by being united, by being, in a sense, "married."

The desire that grips Chomsky and Fodor, then, as it once gripped me, is the desire to turn mentalistic psychology into a science by simply identifying it outright with computational psychology. As I myself once proposed this program (under the name "functionalism"), the way we do this is simple: we simply postulate that propositional attitudes, e.g. *believing that there is milk in the supermarket*, are "functional states" of the brain (i.e. properties defined in terms of program features or other computational parameters, including relations to biologically characterized inputs and outputs).[7] For example, one might postulate that *believing that there is milk in the supermarket* is displaying any formula in a certain computationally defined equivalence class of formulas in a certain way (perhaps with a special symbol for being accepted or asserted, such as Frege's "turnstile," in front of it[8]). Displaying another formula (or any member of another computationally defined class of formulas) in another way (with another special symbol

"written" in front) could be *desiring milk for tomorrow morning's breakfast.*
And going from these two computational states to the resultant, which is the
action of going to the supermarket to buy the milk, might be the result of
carrying out a certain algorithmic procedure on those displayed formulas (as
well as on others). In such a picture, ordinary language mentalistic
psychology, "folk psychology," is a rough approximation to the
computational description of what goes on in the brain. An ideal mentalistic
psychology would literally have a model in the computational structure of
the brain. Make that assumption, and you're off! You have "cognitive
science."

"Cognitive science" is just the latest form taken by a more general
tendency in the history of thought, the tendency to think of meanings (or
"contents," to use a term Fodor favors in his more recent writing) as isolable
and psychologically explanatory ("psychologically real") entities in the mind
or brain. And it is this entire tendency — a tendency I have elsewhere[9]
referred to as "sophisticated mentalism" — that, I shall argue, is
fundamentally misguided.

Three Reasons Why Sophisticated Mentalism Cannot Be Right

I. *Meaning is Holistic*

The doctrine of meaning holism arose as a reaction to phenomenalism, but
the arguments refute any attempt to single out some small subset of our total
vocabulary (say, the observation terms), by means of which it will be possible
to define explicitly all the other terms we use or understand. The arguments I
refer to are associated with the work of W.V. Quine.[10] These arguments are
accepted by Fodor,[11] but I do not think that he has perceived the
singnificance of the holistic character of meaning for what he is doing.

Holism can be understood by contrasting it to two different ways of
looking at language. Holism is, in the first instance, opposed to
reductionism. The reductionist view of language insists that all meaningful
nonlogical words in our language must have definitions in terms of words in
a "basic" vocabulary, a vocabulary which consists of words which stand for
notions which are epistemologically more primitive than the words which
typically occur in, say, theoretical science. The favorite candidate of
reductionists is a vocabulary which consists of sensation terms, or, at any
rate, terms for what is supposed to be "observable." If we formulate
reductionism as a thesis about the truth conditions for sentences rather than
as a thesis about the definability of terms, we may say that, historically,

reductionists have always insisted that the meaning of a sentence should be given (or be capable of being given) by a rule which determines exactly in which experiential situations the sentence is assertible.

Now, most of twentieth-century philosophy of science consisted in the gradual overthrow of this reductionist view. The Logical Positivists themselves shifted from advocating the reductionist view to criticizing it. Basically, what was pointed out was that theories cannot be tested sentence by sentence. If the sentences of which a theory consists had their own independent experiential meanings, or made their own independent claim as to what experience should be like, then one could test a scientific theory by testing sentence 1 and testing sentence 2 and testing sentence 3, and so on. If all of these sentences turned out to be separately true, then the theory would have turned out to be true. But, in fact, the isolated statements of a theory generally have no (or very few) experiential consequences at all. For example, Newton's Theory of Universal Gravitation (without any added statements specifying boundary conditions) is compatible with any orbits whatsoever. (One might even reconcile *square* orbits with the Theory of Universal Gravitation by saying: "Well, that means there are non-gravitational forces acting on the system.") It is only in the presence of a large body of statements that one derives all of the usual "consequences" from a scientific theory. As Quine put it, sentences meet the test of experience "as a corporate body," and not one by one.

The same thing is true of the language of daily life. If I tell you, for example, that *the thief entered the room by that window*, and *there is muddy ground outside the window*, you will "deduce" that *there must be footprints in the mud*. But this is not a deduction (in the logical sense of "deduction") from the just mentioned and only the just mentioned, for you will obviously have employed an auxiliary hypothesis to the effect that *if the thief entered through that window he walked through the mud*, and further items of general knowledge as well. If I say: "No, he was wearing stilts," then instead of expecting to find shoe prints in the mud, you will now expect to find holes of a different shape. What has "experiential import" is the "corporate body" of statements, and this is not the additive sum of the experiential imports of the individual statements.

In ordinary language as opposed to formalized language, this phenomenon is made even more pervasive by what is sometimes called the "non-monotonicity" of the logic of everyday discourse. In a formalized language, if I say "All birds fly" and I also say "Ostriches are birds," one can deduce "Ostriches fly." But in ordinary language it isn't like that. If I say "Hawks fly," I do not intend my hearer to deduce that a hawk with a broken wing will fly. What we expect depends on a whole network of beliefs; if language describes experience, it does so as a network, not sentence by sentence.

Meaning holism has another aspect, which is the following: in addition to representing a thrust counter to reductionism and operationism, it also represents a thrust counter to the great tendency to stress *definition* as the means by which the meaning of words is to be explained or fixed, a thrust counter to that famous stumper "Define your terms!" It has this aspect (which is the aspect most stressed by Quine), because a suggestion that at once emerges from holism is that most terms cannot be defined — or, at least, cannot be defined if by a "definition" one means something that is fixed once and for all, something that absolutely captures the meaning of the term.

Why does holism have this aspect? Because, when an entire network of beliefs runs up against recalcitrant experiences, "revision can strike anywhere," as Quine has put it. Even when a term is originally introduced into science via an explicit definition (imagine that 'momentum' was originally equated with 'mass times velocity' *by definition*), the status of the resulting "definitional" truth or putative truth ("momentum is mass times velocity") is not forever a privileged one. When Einstein discovered that if momentum is a conserved quantity and the Law of Special Relativity is correct, then momentum could not exactly equal mass times velocity, he simply decided that that statement was only approximately correct. The very statement that was originally "the definition" of momentum was the one that was revised! And reasonably so: for why should not the statement that *momentum is conserved* have at least as great a right to be preserved when a conflict is discovered as the statement "momentum is mass times velocity"?

A philosopher of a traditional bent might have answered this last question by saying: "Because "Momentum is mass times velocity" gives the very *meaning* 'momentum'. You cannot revise an *analytic* truth." But such a philosopher is imposing a set of categories (the ideas of fixed meanings of terms and analytic truths) which have no reality for actual scientific investigation. In effect, he treats an accident of linguistics — how the term 'momentum' first came into science — as if it constrained the future choices scientists could make. As Quine puts it, truth-by-stipulation or truth-by-definition is not an enduring trait of sentences. When the statements in our network have to be modified, we have trade-offs to make; and what the best trade-off is in a given context cannot be determined by consulting the historic "definitions" of terms. (Nor can one say: "Well, the scientists decided to *change the meaning of 'momentum'*," for that would suggest we are now talking about a different quantity. No, we are still talking about the same good old momentum — the quantity that is conserved in elastic collisions, the vector quantity that is in the direction of motion of the particle, the one that is *approximately* equal to mass times velocity in the limit of velocities small in comparison to the speed of light. That's the quantity the word 'momentum' always referred to, if it referred to anything. And that quantity, momentum itself, turned out *not* to be exactly equal to mass times velocity.)

In sum, the reasons for the impossibility of explicitly defining most of the terms we use in science or in "ordinary" language have to do with the revisability of our statements. Just as we have come to think that there is no property (unless it be "being Hilary Putnam") which is conceptually necessary and sufficient for me to keep in order to remain the person I am, so there is no property which it is conceptually necessary and sufficient for the referent of a natural kind term to have in order to be a referent of *that* term. Meanings, like persons, are historic entities. Very much as happens in the case of persons, there are practices — social practices — by which we decide when there is enough continuity through change to justify saying that the meaning still exists. Meanings have an identity through time but not essence.

II. *The Central Notion of Meaning Theory — the Notion of Meaning — is Closely Tied to Epistemic and Normative Notions (Notions of Justification, Confirmation, Warrant, etc.)*
I have argued in a series of books and essays[12] that the notions of being a *justified* or *reasonable* or *warranted* or *true* belief are not reducible to physicalistic notions. I cannot review the arguments here. But even if one *could* give a reductive analysis of the normative notion of a justified belief, say, by identifying "being justified" with "being the outcome of such-and-such methods," or such-and-such an algorithm, or such-and-such a computer program, that algorithm would have to be as complex as the description of the brain of an idealized scientist (as complex as the brain of Carnap's "ideal inductive judge"). The discussion of meaning holism just reviewed indicates some of the complications involved: we have learned that testing a scientific theory is not something that can be done in a mechanical way (e.g. by just looking up the operational definition of each term, and seeing how to test the sentences that comprise the theory one by one). Rather, testing a scientific theory involves weighing very intangible things, such as simplicity (which is, in any case, not a single "factor," but different things in different situations), and weighing simplicity against our desire for successful prediction and also against our desire to preserve a certain amount of past

doctrine. It involves having a "nose" for when a trade-off between such desiderata has been made in a reasonable way. The ability to do all this, Fodor calls 'general intelligence,' and he does not expect general intelligence to be explained in terms of "modules" in the forseeable future, if ever. "General intelligence" is a hopeless problem, according to Fodor, and the whole point of the "modularity hypothesis" is to separate the problem of understanding the "language organ" from the presently hopeless problem of understanding general intelligence.[13]

Now, I want to say that the notion of "meaning," or rather the notions (on which the very notion of meaning depends) of "sameness of meaning" and "sameness of reference" have the very same intrinsic complexity as do the notions connected with general intelligence, the notions of "justification" and "reasonableness." But why, the reader may well ask, should it require so much intelligence to tell when two terms have the same meaning? Admittedly, there are cases in which it doesn't. But sometimes it doesn't require much intelligence to solve a scientific problem either. The question of interest is: "How hard can it become in the hardest case?"

Well, at the very least, a theory of synonymy has to decide questions of interpretation. Consider, however, just how subtle questions of interpretation can be. The decision that scientists who used the word 'momentum' intended it to function as a "rigid designator," i.e., as a *name* for a particular magnitude (the one which is in the direction of motion of the particle, is conserved in elastic collisions, etc.), rather than as a synonym for the phrase "mass times velocity" (even if they said that that is the "definition of momentum") has already been mentioned. Another example that I have often used is the decision (which is generally made quite automatically and unconsciously) that when Bohr used the word 'electron' (*Elektron*) in 1934 he was talking about the very objects he called "electrons" in 1900. Note that this is not a case of looking and seeing that the theories which Bohr held in connection with the term at these two times were the same, for they weren't the same. The 1900 theory said that electrons go around the nucleus just as planets go around the sun, i.e. electrons have trajectories, while the 1934 theory (which is, in essence, the present quantum theory) says that an electron never has a trajectory; in fact it never has a position and a momentum at the same time. Yet the history of science is told in this way: we discovered that electrons have a certain charge-mass ratio (by deflecting electron beams in a magnetic field); later we discovered what the electron charge is; we discovered that an electric current is a stream of electrons; we discovered that every hydrogen atom has one electron and one proton; we thought that electons had trajectories, but then we discovered the Principle of Complementarity; etc. In short, we tell the story as a story of continuous change of belief, not as a story of successive meaning changes.

This illustrates what has been called the Principle of Charity or Principle of Benefit of the Doubt[14] in interpetation. All interpretation depends on "charity," because whenever we interpret anyone we have to discount certain differences in belief. Suppose, for example, we are reading a novel written two or three hundred years ago, and we encounter the word 'plant'. We do not even hesitate to interpret this as meaning "plant"; yet, in so doing, we are ignoring a host of differences in belief. For example, we believe that plants contain chlorophyll, we know about photosynthesis and about the carbon dioxide–oxygen cycle, and so on, and all of this was unknown two hundred years ago. Yet we do not follow Kuhn[15] in saying that people two hundred years ago "lived in a different world," or that their notions are "incommensurable" with any notions we now have, which taken *literally* (or course, it never is!) would imply that we couldn't translate a single letter that anyone wrote two hundred years ago. In short, we treat the word 'plant' as having an identity through time but no essence, and we treat the word 'electron' as having an identity through time but no essence.

Yet, we do not always interpret our past words in the way that maximizes the number of true beliefs that people in the past would have had if the interpretation were correct, contrary to some of the cruder versions of the principle of "charity in interpretation." For example, the metallurgist Cyril Stanley Smith once proposed that we ought to say that there really was such a thing as phlogiston (phlogiston, according to Smith, is *valence electrons*). But, in fact, we do not speak as Smith proposes we should; we do not say: "Phlogiston chemists were partly right. They were talking about valence electrons, but they had some of the properties wrong." That would be excessive "charity." The knowledge (whether it be in linguistics or in "real life") that one thing is reasonable charity while another thing would be excessive charity exhibits our full powers of understanding. There is no reasonable hope of a theory of synonymy which is independent of a full theory of general intelligence.

If we refect on the role played by the notion of sameness of meaning in logic, then perhaps it will not seem so surprising that this notion seems to presuppose such complex epistemic notions as the notion of "reasonableness." In logic, *equivocating*, i.e. using a term in one sense at one point in an argument and in a different sense at a different point in an argument, is a fallacy whether the argument is supposed to be deductive, inductive, or abductive (theory-constructive). But the notion of "sense," or "meaning" (Fodor's 'content'), could not play this role in criticism if we did not interpret one another in such a way that "meanings" are regarded as invariant under all but the most unusual procedures of belief fixation. If we adopted the meaning proposals of positivists and operationists, according to which the adoption of a new scientific theory is *always* a change in the "meaning" of the theoretical terms involved, then we would simply have to invent a new notion of "change of meaning" to distinguish between the case in which "You have changed the meaning of the terms involved in my original question!" is a legitimate criticism and the case in which (on the positivist theory of "change of meaning") it isn't a real criticism (because one *had to* invent a new theory to answer the original question). My view is that we interpret one another so that sentences remain synonymous with themselves under normal procedures of belief fixation. It is the fact that interpretative practice owes allegiance to this constraint that, in large part, accounts for the fact that sameness and difference of meaning fail to coincide with the presence and absence of any isolable computational relation among our "mental representations." Certainly, as our examples illustrate, such a relation could not be "modular," that is, it could not be psychologically more elementary than full scientific intelligence.

III.   *Meanings Depend on our Physical and Social Environment in a Way that Evolution (Which was Over, for our Brains, about 30,000 Years Ago) Couldn't Foresee*
To have given us an innate stock of notions which includes *carburetor, bureaucrat, Original Sin, quantum potential*, etc., evolution would have had to be able to anticipate all the contingencies of *future* physical and cultural environments. Obviously it didn't and couldn't do this.

Connections Between I, II, and III

These three points are interconnected in ways it is important to see. The argument against reductionism (say, phenomenalism or operationism) and against the possibility of defining all of our concepts explicitly from some basic stock was summarized under point I. The heart of the argument was that to adopt a notion of "meaning," according to which ordinary scientific discoveries (discovering that water is H_2O, that momentum is not exactly mv, that electrons obey the Principle of Complementarity, or that plants contain chlorophyll and perform photosynthesis) *change the meaning* of the relevant terms, would violate a principle mentioned under point II, the principle that meanings are invariant under ordinary processes of belief fixation. To say that we changed the meaning of the word 'water' when we decided that water is H_2O would not ony violate the ordinary use of the words 'change the meaning' (or, as linguists prefer to put it, "go against our intuitions of synonymy"); it would violate this principle, which is central to the *epistemic* function of the notion of "change of meaning."

Similarly, there is a connection between points I and III: if reductionism were true (i.e. if point I were false), then evolution would not *have to* give us implausible "innate" concepts such as *positive charge* or *ego cathexis*, even if the "innateness hypothesis" were true; it would only have to give us some stock of basic notions (say, the observation terms) from which they could be defined. But (as Fodor and Chomsky recognize) our terms cannot be defined explicitly from a set of terms much smaller and biologically more primitive than the whole lexicon itself. It short, the truth of *meaning holism* blocks the only way of meeting objection III that makes biological sense. (In *The Language of Thought*, Fodor does not really try to answer objection III; he simply marvels at the fact that all these concepts *must be* innate — since that is required by the facts, on his theory — *in spite of* such objections as III.) In sum, sophisticated mentalism of the M.I.T. variety is not blocked by any one of these points separately, but by I, II, and III acting together.

An Example

Since the discussion has been on an abstract level, at this point I propose to discuss the "Ruritanian" example from "Computational Psychology and Interpretation Theory."[16] In the example, one of the differences between the dialect of Ruritanian which is spoken in the north and the dialect spoken in the south is that in north Ruritanian 'grug' means "silver," while in south Ruritanian this word means "aluminum." We are supposed to imagine that silver is so common in north Ruritania that in the north the pots and pans are made of silver. One might imagine that in the Middle Ages 'grug' meant "pot metal" in Ruritanian, and that it is the fact that north Ruritanian pots are made of silver and south Ruritanian pots of aluminum that accounts for the meaning shifts that have taken place. In any case, northern children grow up knowing that pots and pans are normally made of 'grug' and southern children grow up knowing that pots and pans are normally made of 'grug'.

In the example, Oscar and Elmer were supposed to be in the same psychological condition (in Fodor's "solipsistic" sense of "psychological condition"), i.e. the same with respect to all internal parameters relevant to language at t_0.

So, although in the adult communities to which they belong, 'grug' has two different meanings, it has (in Fodor's sense) the same *content*[17] for Oscar and for Elmer at t_0. At t_1 (when they have become adults) the words must differ in content in the two idiolects as much as 'silver' and 'aluminum' do for speakers of English. Hence the word must *change content* between t_0 and t_1 (for at least one of the children, and presumably for both).

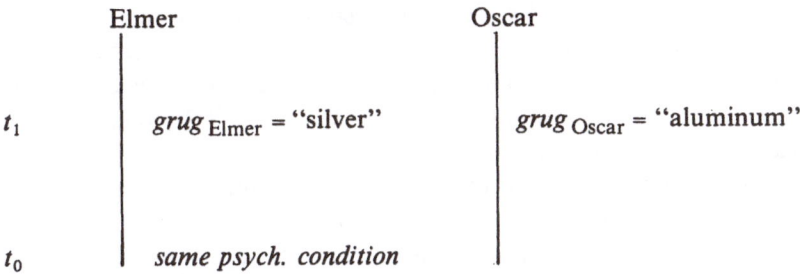

Elmer   Oscar

t_1 *grug* $_{\text{Elmer}}$ = "silver"   *grug* $_{\text{Oscar}}$ = "aluminum"

t_0 *same psych. condition*

However, all that happened between t_0 and t_1 was normal belief fixation (in the model commonly employed in inductive logic, conditionalization of prior probabilities to new information as it continuously pours in). At t_0 both children know that 'grug' is a metal, that it is shiny-gray in color, that it tarnishes, that Mother has 'grug' pots and pans, etc. By t_1 Oscar knows that 'grug' is called 'aluminum' in American English and "aluminium" in British English; that it was briefly very expensive (Napoleon had aluminum jewelry, and it was more costly than platinum), but became very cheap; that pots and pans are normally made of 'grug' or of steel or of copper except in north Ruritania, where silver (called 'zilber' in south Ruritanian) is used for pots and pans; that 'grug' comes from bauxite (they really teach children a great deal in Ruritanian schools!); that *zilber* (which is called 'grug' in the northern dialect) is an expensive metal which does *not* come from bauxite; etc.; and Elmer knows that 'grug' is called 'silver' in English; that 'grug' is an expensive metal; that 'grug' does not come from bauxite; etc. From the "internal" point of view, Oscar would not feel that the acquisition of any of this information was anything but the acquisition of ordinary factual information involving a notion he already had, the acquisition of further facts about 'grug'. A theory according to which the word 'grug' changed its *content* in the course of this perfectly ordinary process of belief fixation flies in the face of fundamental properties of the notion of meaning. Moreover, any place we decide to stipulate that the difference in "content" has taken place will be quite arbitrary, and unrelated to actual practices of paraphrase and interpretation.

In addition, if we postulate that the word 'grug' is attached to a formula in "mentalese" (a "semantic representation") at t_0 (in both brains, since the children are in the same psychological condition), then if normal processes of belief fixation do not alter the "semantic representations" of words, the mentalese formula — say, XYZ — will end up having *two different meanings* at t_1. Pushing the problem of meaning back to "mentalese" will have solved nothing, for we will need a theory of meaning of "mentalese" just as much as we need a theory of meaning for any natural language. (In his recent writing Fodor has abandoned the idea that the "contents" are identical with formulas or even with computationally defined equivalence classes of formulas, in the "language of thought." But now it becomes totally mysterious what these entities are supposed to be.)

These problems do not arise for my own view of meaning, because I regard meaning as a social construct, not as something "psychologically real." In my view, it is legitimate — in fact, it is an essential part of the practice on which meaning-notions rest — to attribute the meaning and reference a word has in their linguistic community to individual speakers, including children who are to some extent "plugged in" to the social network, even if they are not in a position to fix the reference by themselves. On this theory, the meaning of 'grug' is different for Oscar and for Elmer even at t_0. In fact, it means 'silver' in Elmer's idiolect and 'aluminum' in Oscar's idiolect. The meanings can be different even though the "psychological condition" is the same. Meanings aren't in the head. My theory is not a *mentalistic* one.

The theory I presented in "The Meaning of "Meaning""[18] did have a mentalistic aspect, however, and this I would now give up. In that essay I suggested that we take the psychological component of meaning (the "content," in Fodor's terminology) to be the stereotype associated with the term (or the stereotype together with the syntactic and semantic "markets"). It now seems clear to me that we do not and should not require *sameness* of stereotype in interpretation, but rather sufficient similarity, where what counts as sufficient similarity is highly context-sensitive. In short, I would give up the idea that there is an "object" which is the "psychological component" of a "meaning."

What we have are not biological objects which can be identified with "meanings," but a practice of equating terms and statements which is context-dependent and not controlled by a "rule" which it would be easier to formalize than general intelligence.

My Present Picture

I shall close by describing the bearing all this has on the question of "functionalism." The functionalist picture was a two-level picture,

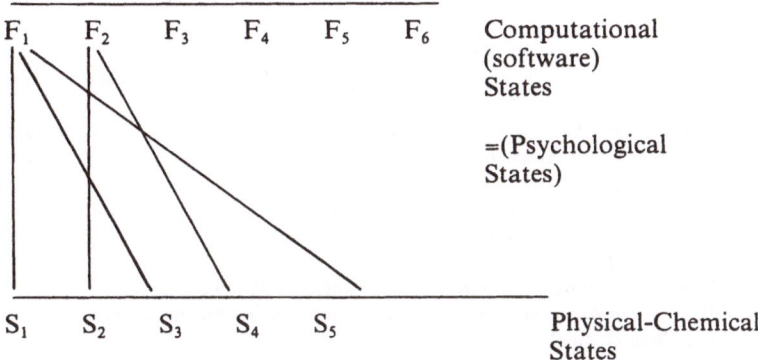

in which each psychological state was identified with a computational state which could be realized by a large (potentially, an infinite) number of physical-chemical states. Later it was realized that not only could a given computational state be realized in infinitely many disparate ways whose relationship to one another could not be seen at the physical-chemical level, but that one physical state could also be the realization of an infinite number of different computational states; that a physical-chemical state does not have an "intrinsic" computational description. Thus the relationship of the states on the two different levels is many-many.

This was an *anti-reductionist* picture insofar as it denied any direct reducibility of psychological properties to physical-chemical properties. But there was still supposed to be a reducibility of psychological properties to computational ones. Today, in view of the sorts of considerations just reviewed, I would move to a three-level picture:

——————————————————— Psychological States
(believing, desiring,
thinking that, etc.)
——————————————————— Computational States

——————————————————— Physical-Chemical States

—with many-many relations between any two levels. In all this, what we are gradually seeing is a breakup of the reductionist picture.

Notes

1   *Language and Thought* (New York: Thomas Y. Crowell Company, 1975).
2   See, for example, Chomsky's *Reflections on Language*, Chapter 1 (New York: Pantheon, 1975).
3   Chomsky speaks of "a subsystem (for language) which has a specific integrated character and which is in effect the genetic program for a specific organ" in the discussion with Piaget, Pappert and others printed in: *Language and Learning*, ed. Massimo Piattelli (Cambridge, Mass.: Harvard University Press, 1980). See also the reference in n. 2.
4   See Chapter 5 of my *Mind, Language and Reality* (*Philosophical Papers*, vol. 2, Cambridge University Press, 1975).
5   Review of Skinner's *Verbal Behavior*, *Language* 35 (1955): 26–58.
6   *Problems of Knowledge and Freedom* (New York: Pantheon, 1971).
7   See Chapter 21 of the book cited in n. 4.
8   Of course, the brain's "language," if it exists, is not literally *written*. See my "What is Innate and Why," Chapter 14 of the Piattelli volume cited in n. 3.
9   In "Meaning Holism," forthcoming in: *The Philosophy of W.V. Quine* in the *Library of Living Philosophers* (Carbondale: University of Southern Illinois).
10 "Two Dogmas of Empiricism" (originally published in the *Philosophical Review* in January 1951), reprinted in Quine's *From a Logical Point of View* (Cambridge, Mass.: Harvard University Press, 1961); "Carnap on Logical Truth," originally published in *The Philosophy of Rudolf Carnap* (LaSalle: Open Court, 1963), reprinted in *Quine's Ways of Paradox* (2nd edition, Harvard, 1976); *Word and Object* (Cambridge, Mass.: M.I.T. Press, 1960).
11 See, for example, his *Psychological Explanation* (New York: Random House, 1968).
12 *Reason, Truth and History* (Cambridge University Press, 1981); the papers in *Realism and Reason* (vol. 3 of my *Philosophical Papers*, Cambridge University Press, 1983).
13 See *The Modularity of Mind* (Cambridge, Mass.: M.I.T. Press, 1983).
14 See Chapter 13 of the book cited in n. 4.
15 *The Structure of Scientific Revolutions*, 2nd edition, enlarged (Chicago: The University of Chicago Press, 1970).
16 See Chapter 8 of my *Realism and Reason*.
17 See Fodor's "Cognitive Science and the Twin-Earth Problem," *The Notre Dame Journal of Formal Logic* 23 (April 1982): 98–118.
18 Chapter 12 of my *Mind, Language and Reality*.

Meaning and Our Mental Life
A Comment

It is altogether impossible for me to comment on all, or even on most, of the issues brought into focus and woven together so masterfully by Professor Putnam. Having been a close Putnam-watcher for quite a few years, I cannot cease to marvel at how Putnam takes, say, a mere technical point in logic, or a seemingly simple story (such as the story of the two Ruritanian children), and uses it like a magician's wand to invoke a whole host of fundamental problems in philosophy. Then, by twirling his wand — that plain example he uses — he forces those philosophical mammoths to change, transforming them in front of our gaping eyes. I think we have all learned a lot today. Before I say anything else, I would like, then, to thank Professor Putnam for the rare intellectual experience he gave us.

I shall now try to discuss briefly two themes from Putnam's paper: the one is his famous contention that "Meanings are not in the head," the other, the question whether there are "in the head" such entities as intentions, desires, and beliefs.

I

For many a year Putnam has insisted on the severe limitations of the Carnapian, verification-plus-probability model of using language. This

33

E. Ullmann-Margalit (ed.), The Kaleidoscope of Science, 33–37.
© 1986 by D. Reidel Publishing Company.

model would account for our understanding a language in terms of (1) language entry rules, (2) assignment of probabilities to registered sentences, (3) inductive and deductive intra-language moves, (4) application of utility functions, and finally (5) language exit rules. But, while in his previous period Putnam sought to supplement this model by Metaphysical Realism (intended to show how such a language can capture *truth*), in recent years he aspires to supplement the currently fashionable mentalistic version of the said verificationist semantics by a set of *many* holistic, interest-relative, value-laden, radical interpretations. It is only on such a radical interpretation, Putnam now claims, that our mental life can be seen as a rational network of concepts and beliefs.

Essential to this project is the proof that meanings cannot be accounted for either by traditional verificationist semantics or by its new, cognitive psychologists' version of localized mentalistic representations. That is, if the project of radical interpretation is to be shown necessary for our understanding of language, then one ought to show that (using the catching Putnamian phrase) meanings are not in the head. This is what the ingenious story of Oscar, the North Ruritanian, and Elmer, the South Ruritanian, is designed to do.[1] The structure of Putnam's argument, I think, is as follows: (1) By hypothesis, 'grug' in North Ruritanian has a different reference than 'grug' in South Ruritanian. (2) When Oscar and Elmer are adults they clearly use 'grug' in different senses; one translates 'grug' into the English word "silver," the other translates it as "aluminum." (3) At no time during their lives, however, did 'grug' change its meaning for either Oscar or Elmer; as Putnam puts it, *learning*, i.e. the conditionalization of prior probabilities to added information, cannot be a change of *meaning*, or else all our reasoning would be flawed on the charge of gross equivocation. (4) Therefore, even as children, Elmer and Oscar used 'grug' in different senses, i.e. they had different concepts of 'grug'. (5) But, as children, Elmer and Oscar had the *same* beliefs about 'grug'; to put it more exactly, the set of all mental sentences including the word 'grug' in Elmer's head was, at that time, strictly identical with the set of all mental sentences including 'grug' in Oscar's head. (6) Therefore, identity of mental contents does not determine identity of sense; *ergo*, meanings are not in the head.

Can people such as myself, who believe that there is no other place for meanings to be in except in the head, rebut this argument? Steps (1) and (2) are true by hypothesis. Step (3) is the crucial one, but, unlike Feyerabend, etc., I believe it is above reproach; I am convinced that Putnam's reasoning here is compelling. So I do not have a choice but to accept (4); I shall not go into the matter here, but I also have some other reasons for adopting (4). It is step (5), then, which I take exception to. I shall try to show that (5) is ambiguous between two readings, and the one needed to entail (6) is false.

The set of the mental sentences including 'grug' in the heads of the two children is *identical*. The set of propositions maintained by the two children is *not* identical. So far, Putnam and I agree. How to account for the divergence? Putnam places the blame on the meaning of 'grug'. I don't. We must note that the 'grug'-including sentences in the children's heads include *indexical* words. Elmer says: "*These* pots and pans are made of 'grug'" and so does Oscar; but upon saying 'These' they point at different pots and pans. Oscar says: "*Grug* is the stuff *my mother* calls 'grug'," and so does Elmer, but Oscar and Elmer have different mothers. Borrowing Kaplan's distinction between senses (world-to-reference functions) and characters (context-to-sense functions), we can say that 'my mother' has the same *character* for both Elmer and Oscar, but different *senses.*

Putnam, of course, may be quite happy with this formulation, since he can claim, as Kaplan in fact does claim, that the element which assigns different senses to the same character is the actual reference of the indexical words, and this is surely not a psychological element. The *word* 'grug' may literally be in the children's heads, but its reference, the *metal grug*, is surely not to be found in the heads of human children! Kaplan's conclusion, which suits Putnam perfectly, is that although the *characters* of indexical expressions are in the head, their *senses* are not. Thus the sense of 'you', for example, includes you, but *you* are not in anyone's head. So then, the question whether senses are or are not in the head boils down to the question whether the argument, which takes characters into senses, is or is not a psychological element. My own view is that it *is* psychological. In a paper named "*De Se* and Descartes," forthcoming in *Nous*, I propose a semantics for indexicals based upon the idea that the word 'I', for example, has a different *sense* on each occasion of its use. That sense is determined by the constant *character* of 'I', which Putnam and Kaplan may admit is a psychological entity, and by the user's *experience* at the time he or she uses 'I', the latter also being a psychological entity. I therefore maintain that the *sense* of 'my mother' is different for Oscar and for Elmer *not* because they have numerically different *mothers* (which is a non-psychological element) but because they have numerically different *experiences* (which is a strictly psychological element). Thus, not only indexical characters but also indexical senses are in the head. I would therefore agree with Putnam that when Elmer says, for example, "*This* pot is grug," the sense of what he says is different from the sense of what Oscar says in uttering the same sentence. But the reason for this difference, I think, has nothing to do with readiness to attribute different properties to 'grug', which, as Putnam has clearly shown, may or may not exist. Rather, it is a function of the different senses of the indexicals used by Oscar and Elmer.[2] If I am right in maintaining that these senses, too, are psychological entities, then there is no reason not to believe that meanings are all in the head.

II

The other point I wish to discuss, very briefly, is Putnam's contention (which he shares with Donald Davidson) that the intentions, beliefs, desires, preferences, and other psychological entities which the best psychological theory may ascribe to a person on the basis of a rational reconstruction of that person's behavior need not be encoded in the brain. The set of all mental sentences which in fact represents that person's attitudes may be just as inchoate and self-contradictory as the set of natural language sentences evinced, in good faith, by that subject.

Those who do not appreciate the force of this argument may raise here a well-known objection which I think Putnam can easily negotiate. The objection is this: Isn't Putnam's point merely the old argument of metaphysical realists, that the world may *really* be totally unlike what the best theory we have (or shall have) takes it to be? Putnam rejects the latter kind of skepticism as meaningless (I, as a metaphysical realist, do not agree with Putnam on this point). So why not take the same attitude with respect to skepticism concerning the truth of theories in cognitive psychology?

Putnam may answer, I think, that the case of psychology is quite different. Theoretical models in science are bound by methodological constraints, but one is not allowed to stipulate what kind of entities the best theory of nature will come up with. Cognitive psychology, on the other hand, is supposed to explain behavior, not only by providing a *causal* explanation of that behavior, but also by showing how this behavior constitutes *rational action*. The internal mechanism which brings about that behavior is to be that of a *rational* and *moral* agent. This is the famous Principle of Charity which, as Davidson has shown, must be presupposed by all explanations of behavior in terms of desires, intentions, and beliefs. If this constraint is removed, there is no reason why the elements of one's psychology should stand, e.g. in the *logical* relations manifested by the elements of practical syllogism. Surely no one expects protons and electrons to stand in such relations to each other?

I wish to suggest two answers, the first well known, the second, I think, new, to this kind of skepticism. First, it is highly implausible to suppose that a complicated, fruitful, and predictively successful theory can turn out to be that way *by accident*. If there is nothing in the mind which corresponds to the network of intentions, beliefs, and desires by means of which we can successfully account for human behavior, our success is miraculous and quite bewildering, and cross-confirmations of psychological laws, on various occasions, an act of a god or a devil.

My second answer is that, in my opinion, the Principle of Charity is an essential constraint on *all* scientific models. The principles of Coherence, Parsimony, Fecundity, Predictability, are all, *au fond*, aesthetic principles: they require that a good theory be elegant *as a theory*. The Principle of Charity is different: it, too, is an aesthetic principle, but what it says is that a theory, T, is preferable to rival theories if the object described by all these theories has a highest value as described by T. Indeed, we use this principle as a matter of course in our understanding of persons and in all commonsense explanations. But it is also used in science, under the misleading name, *'Simplicity'*. For example, if we were to rewrite the laws of mechanics, attributing slightly irregular trajectories to moving bodies, this would not change the theory's predictive power, its explanatory power, its ontological and logical commitments, its internal coherence, its amenability to integration with other theories in other fields, etc. If we are accustomed to calculate such magnitudes, we may even find the new system pragmatically simpler. What, however, will be lost is the *beauty*, not of the theory, but of the *world*. The world, as described by that theory, is ugly. It is then the value of the *object* which, *via* the Principle of Charity, enjoins the acceptance of the laws which we do in fact accept in physics. But if Charity is prescribed to the physicist, surely the psychologist may be allowed to practice it, too?

Notes

1   This example is far better than Putnam's previous example of the person who cannot tell a birch from an elm. In that case, it is not true that the said person would have the same beliefs about birch and elm. Even if one has no idea what kind of tree birch (or, elm) is except that it is a kind of tree, one would still know that birch has one property which elm does not have, i.e. that it is called 'birch' in English. (In our culture this information is most important, since, by making use of it, one can find other differences between birch and elm, e.g. by consulting a dictionary.)

2   To be a bit more precise, my view is that the character of the type-word 'I' is the following mental open sentence: [The experiencer of x]. (I use square brackets to denote sentences in Mentalese.) To get the *sense* of a given token of 'I', fill the empty argument place in the above open sentence with the experience of uttering that token. But note: the argument is an *experience*, not any name or term (indexical, descriptive, or otherwise) denoting one.

The Persecution of Absolutes:
On the Kantian and Neo-Kantian
Theories of Science

AMOS FUNKENSTEIN

The Theme

Kant's *Critique of Pure Reason* intends, in part, to be a theory of science; to what extent is a matter of interpretation. Of the German philosophical systems inspired by him, some sought to abolish, conserve and transform his premises all at once — in the famous triple sense that Hegel ascribed to 'sublation' (*Aufhebung*). Others regarded their own philosophy as a creative interpretation of Kant's system. The former group, the *speculative idealists* like Fichte, Schelling and Hegel, lost any specific interest in the foundations of the exact sciences — in the same way as scientists rarely sought contact with them: they rather tended to stress the gap between speculation and *Wissenschaft*. A scientific orientation was retained or revived among those philosophers who viewed themselves as authentic interpreters of Kant — from Solomon Maimon to the School of Marburg. Neo-Kantianism, like phenomenology a generation later, began with a protest against vulgar positivism (or psychologism). Yet Husserl — particularly in the *Logische Untersuchungen* — took some of his leading models from pure mathematics and tried to develop an *empirical idealism* of sorts, i.e. a taxonomy of concrete *a priori* entities; while the Neo-Kantians of Marburg (as opposed to those in Baden) were guided by models derived from mathematical physics.

39

E. Ullmann-Margalit (ed.), The Kaleidoscope of Science, 39–63.
© *1986 by D. Reidel Publishing Company.*

With the aid of these models Cohen, Natorp, Kellermann and Cassirer constructed a system of *idealistic empiricism* as a theory of science.

Their theories are all but forgotten today, but unjustly so, because their imprint is still recognizable in recent discussions concerning the criteria of rationality in science and the role of rationality in the history of science. Some of the most original insights of the Neo-Kantians have now virtually become truisms. Hermann Cohen, long before Popper, Fleck and Feyerabend, destroyed the artificial fence between 'facts' and 'theories': every fact, he insisted, is theory-laden, and yesterday's theory may be today's fact. Scientific reasoning cannot be abstracted from its sources and context. As a result, for the first time since the Peripatos, logic was again defined as a system of rules of inference of scientific propositions (*Urteile*) embodied in concrete scientific disciplines, rather than as the rules of right reasoning or of formal reasoning. In other words: To the Neo-Kantians, logic became once again a mere instrument (*Organon*) of concrete sciences. For similar reasons, they were also the first to see in the history of science an essential foundation to any theory of science, rather than a pool of corroborating (and ultimately contingent) examples. A true history of science, they believed, is the only basis for constructing a theory of science, contrary to our inclination to measure the history of science against an abstract theory. Science, as viewed by the School of Marburg, does not derive its validity and vitality from principles outside itself. It creates of itself its methods and objects, facts, theories, and the criteria of discrimination between theories and facts.

Such and similar positions certainly have a ring of modernity to them. Yet they deserve our attention not merely out of a sense of gratitude, because they were the first, and because the traces of their influence had become almost obliterated; but because their arguments can still contribute to our present disputes. If we take the trouble to lift the terminological veil which separates these arguments from ours, we shall find them relevant and fruitful. The Neo-Kantians' 'Principle of Origin' (*Ursprungsprinzip*), so I shall argue, is an interesting variant of dialectical logic. Contrary to Hegel's dialectics, which is circular and closed, it allows the system of science to be open and can therefore serve as a powerful instrument to elucidate the dynamics of change of scientific theories. Furthermore, the historical career of the Neo-Kantian movement, its achievements and failures, may teach us something about the potentialities and limits of any systematic philosophy of science.

Our study has therefore both a systematic and a historical dimension. Firstly we shall examine the historical sources of Kant's theory of science. The seventeenth century, as I argued elsewhere, united two ideals which had previously existed separately: the ideal of an unequivocal scientific language and the ideal of the homogeneity and uniformity of nature. The new merger expressed itself in a new notion of the laws of nature, of physical necessities; in the seventeenth century it was still justified by means of theological arguments, and advanced a methodical concept of God as a warranty for the intelligibility of nature. Kant's theory of science, we shall further argue, was a radical departure from the methodical God; Hermann Cohen concluded this process of detheologization of science with his attempt to ground its validity in itself only.

Thereafter we shall examine Cohen's system in terms of its time. Cohen epitomized, no less than the most enthusiastic positivists, the *Wissenschaftsgläubigheit* of his period, the belief in the steady, organic growth of scientific knowledge under one canon of rationality. In many ways his system was already obsolete when formulated. Cohen saw analysis as the culmination of mathematics — a view fitting, perhaps, for the beginning of his century, but no longer for his time. Even in the study of nature, other mathematical branches, e.g. matrix and group theory, became later no less important than differential equations. And Cohen, like Kant, saw in classical mechanics a paradigm of scientific knowledge but failed to notice the already visible signs of its crisis.

Finally we shall seek to understand why, notwithstanding its shortcomings, the School of Marburg nevertheless succeeded in constructing tools of thought by means of which the crisis of science could be explicated. Perhaps the excellence of philosophical systems is recognizable by their ability to dig their own graves. The unique instruments which they create to defend certain crucial positions prove to be so valuable that they survive the system itself or even help to undermine it. Such was the conception of science as a context in which facts and theories, objects and modalities of their cognition, are interdependent throughout. Such was Cohen's decision, in contrast to Kant, to understand science from its dynamic process and to recognize, at the heart of this process, a principle of *constructive ambiguity*.

Kant and the Detheologization of Science

Kant — and later Cohen — sought to erase the notion of God as a methodological guarantee for our understanding of nature. He reacted against various and continuous efforts in the seventeenth and eighteenth centuries to interpret God as a warranty for, or even as the embodiment of, the rationality and intelligibility of the universe. Why should we assume that nature is well structured, and hence intelligible? In what sense are laws of nature 'universal and necessary'? Their necessity is evidently not logical. The negation, say, of the universal law of gravity does not entail a logical contradiction. A universe is indeed conceivable in which bodies repel (rather than attract) each other in direct proportion to their masses and in inverse proportion to their squared distances; an uncomfortable universe perhaps — but not a self-contradictory one. Should we wish to base 'physical necessities' on the somewhat weaker ground of induction, we are led into paradoxes which obtain even if we admit that laws of nature are never verifiable and at best falsifiable.[1] The very meaning — let alone justification — of law-like statements in science is as dubious today as it has been since the rise of early modern science.

The methodological concept of God — an answer to these or similar doubts — was born with the onset of modern philosophy. Indeed, no figure of thought illustrates the shift from medieval to early modern philosophy better than the manner in which, throughout the seventeenth and eighteenth centuries, the very concept of God was designed to secure or even embody the complete rationality of the universe. True, the language is often medieval; yet for all the medieval terminology in which this 'methodological God' is disguised, the manner in which Descartes or Leibniz made God, God as such, God in his totality rather than any particular attribute of his, vouch for the complete rationality of the world, stands in sharp contrast to the medieval sense of God as, first and foremost, the source of all contingencies — even where he vouches for a certain amount of harmony of the world. Leibniz boasted that he was first to distinguish between the *nécessité logique*, which is grounded on the principle of noncontradiction and under which even God's power is subsumed, and the *nécessité physique*, which is grounded on the principle of sufficient reason and "inclines God without necessitating him."[2] But he must have known better: this is but another version of the medieval distinction between God's absolute and ordained power (*potentia dei absoluta et ordinata*). The scholastic distinction introduced, for the first time in the history of Western thought, the difference between logical neccessity and natural order. In the later Middle Ages, the schoolmen were driven by an obsessive compulsion to actually devise orders of nature different from the one admittedly in existence, so as to enlarge the horizon of the divine omnipotence. If God so willed it, the Earth would cease to be the center of the universe and the proper place of all heavy things; God could even move the whole universe in a straight line indefinitely in empty space. Had God so wished, the savior of the world could have been a stone or a donkey — *aut lapis, aut asinus.*[3]

The difference between the philosophy of the Middle Ages and that of the early modern era lies not so much in the forms of reasoning they employed, but rather in their application. To the traditional *quaestio scholastica* whether God could have created a world better than ours, Thomas Aquinas and Leibniz gave different answers — albeit both of them were convinced that the universe is indeed harmonious. The God of Thomas Aquinas cannot choose rationally between the infinitely many possible universes he is capable of creating: if he wished to create a world, he had to choose one of them arbitrarily, since instead of every world, good as it may be, God in his omnipotence could have created a better one.[4] There cannot be a world order which is the best. Leibniz' God is also capable of actualizing, if he so wished, any one of the infinitely many possible universes; but since he is rational, the principle of sufficient reason inclines (yet does not necessitate) him to choose the best-ordered universe, i.e. the largest cluster of compossibles subject to the least number of constraints. God guarantees the foundation of laws of nature, the validity of the principle of sufficient reason. He vouches for the complete intelligibility and rationality of the universe, and was thus transformed into a methodological device; the renewed interest in the ontological proof of God's existence since Descartes is a part of this methodological conversion of the concept of God, the *ens necessarium*.

The history of the methodological functions of God in the understanding of nature cannot be elaborated here. I am concerned here only with the decline of this concept, without referring to its finest hours. Kant was perhaps the first to have grasped and articulated this merely methodological function of God within recent metaphysics; his refutation of all proofs of God's existence is based on the demolition of the methodological functions of the concept of God in any future understanding of nature.

Like some of his rationalistic predecessors, Kant recognized that the idea of the *absolute* rationality (or intelligibility) of the 'totality of things' demands, entails the so-called principle of 'complete determination' (*durchgängige Bestimmung*) of *every* thing.[5] Kant's distinction between the logical 'determinability' of a *concept* and the consistent, transcendental 'determination' of a *thing* may be illustrated as follows: A rational number is always completely constructed by two integers. An irrational number, though it can never be completely constructed, is always constructible to any desired precision. To the question: is the nth number after the digit the number four? the answer is always yes *or* no; I can construct the irrational number up to (n) and determine its value. But the irrational as a whole is never completely determined, i.e., completely constructible. Or consider another illustration. To the question whether Napoleon had a Muslim ancestor the answer is already determined as yes or no, even if in practice I may never be able to ascertain it. But the same question asked about Stendhal's Julien Sorel is neither yes nor no until I asked and answered it arbitrarily, because Stendhal was mute on that point. Julien Sorel is *only* a concept because he is not 'completely determined'.

Completely determined means determined against all possible simple predicates, be their number finite or infinite. Simple predicates are those which neither imply nor exclude each other. Leibniz' monads were construed from such simple predicates. Kant called them, *pace* Baumgarten 'realities' rather than 'perfections'. Both terms refer to a long tradition according to which the positive predicates (attributes) of a subject add to its reality: the more positive attributes, the more real it is. As just stated, all of them are compatible by definition; wherefore, Kant says, a hypothetical subject is indeed conceivable in which all simple predicates inhere — and it would then be 'the most real thing', an *ens realissimum*, which embodies the idea of the unity of all realities (perfections) — *der Inbegriff aller Realitäten*. But this concept is nothing but a hypostatized version of the assumption of a complete determination of every 'thing'. Claiming that such a being is conceivable is far from saying that this *ens realissimum* — which our reason 'hypostatized, thereafter personified' — must be conceived as existing, that it is a necessary being. Existence is not a predicate, hence not one of the realities attributable of necessity to the 'most real being'. Kant admits that another concept, the concept of a necessary being (*ens necessarium*), does entail existence: but it is a vacuous concept, a concept without further content. The fallacy of the ontological proof of God's existence, even in its most sophisticated elaboration (such as Leibniz'), is not that it understood existence as an attribute (this is how Kant's refutation is usually rendered). It rather lies in the arbitrary and hence mistaken identification of two concepts of reason — the concept of the most real thing being the concept of the necessary being. The ontological proof is a case of mistaken identification of two ideas of pure reason.

More important to our concerns is the circumstance that Kant set out to prove that the principle of complete determination, and with it the methodological concept of God, is at best a regulative ideal of reason and has no bearing whatsoever on our actual interpretation of nature by our understanding. In a more recent idiom we might say that they are not theoretical, but at best metatheoretical assumptions. They are the means by which pure reason (*Vernunft*) conceives of totalities in themselves and in their ultimate, discrete, but altogether abstract components: things in themselves are conceived as completely determined. But neither the principle of complete determination nor the concept of a 'sum-total of all realities' is a part of our experience of nature (*Erfahrung*). Nor are they necessary to understand experience, to structure experience, as are the categories of understanding. The interpretation of experience does not demand the construction of absolutely simple predicates and of things-in-themselves in which such simple predicates could inhere. Even Leibniz admitted that much. Simple predicates and their sum-total are pure abstractions. The rationality and coherence of our experience — i.e. of nature — and of the spontaneous categories with which we grasp and pattern it are grounded not on the principle of complete determination, but rather on the 'synthetic unity of consciousness'. In other words, the coherence and consistency of our experience do not demand or entail the assumption of the ultimate coherence of the 'totality of all things'. It suffices to assume (as we must) that, *if* an entity existed which has no orderly link to any other members of experience, such an incoherent something could not be perceived; just as a mathematician cannot deny the existence of totally random sequences of numbers, even though he must insist, by definition, that there can be no *formula* to construct such sequences. If there were such a formula, the sequence would *eo ipso* not be a random one. Our experience is *a priori* patterned, coherent experience. This is a much more modest claim than that for the unity and coherence of the 'totality of everything', which is unprovable and perhaps even self-contradictory.

All this is not to say that the methodological concept of God — and the principle of thoroughgoing determination — are not *linked* in some ways to our understanding (*Verstand*). The principle is indeed a projection of, or extrapolation from, a very basic figure of logic from which one important category of understanding is also derived, namely that category which permits us to quantify qualities and to discriminate the real from the unreal. Since this is also the very starting point of the Neo-Kantian theory of science, a more elaborate explication is needed for Kant's distinction between 'negative' and 'infinite' judgments and the category of 'limitation' which corresponds to that distinction.

Lies may lack a leg to stand on, but they have many faces. Among the many complicated problems of negation is also this one: Can a logical system, as formalized as can be, do with only *one* form of negation? The answer appears to be negative; we must, so it seems at first sight, distinguish various modes of negation — so as to distinguish between negation within a statement and the negation of a statement, between meaningless and false propositions, between well-formulated formulas and nondemonstrable formulas, between factual and category-mistakes, and so on. Or should we argue, *pace* Prior and others, that negation is always one and the same, while its causes may be various? A statement may be negated because it is counterfactual, or because it involves a category-mistake, or because it is meaningless, yet in any one of these cases the negation functions in the same manner. The statement "Napoleon won the battle of Waterloo" is false to the same measure and in the same sense as the statement "Admiration is triangular" or even "Within not dances even," though for different reasons. Or again: each of them denies another attribute of a statement (or of its parts) — say factuality, possibility, meaning; but the negation itself has always the same meaning, even if incapable of further definition.

I very much doubt the validity of arguments like this. Yet I refrain from elaborating the point for fear of entering a semantical maze from which there is no return. This much, however, can be proved: some distinction, either between modes of negation, or causes for negation, or any other binary disjunction of negative propositions, is logically necessary. Its necessity does not emanate from usages of language which may result from an erroneous logical intuition; we are led to it by purely formal considerations. Even in a well-formalized system of propositions, if it is to be consistent and rich enough to express at least numerical relations in our world, we cannot reduce negation to one form, or mode, or cause, or interpretation only. As is well known, it has been mathematically proven that every formal system which suffices to derive the propositions of arithmetics from a finite subgroup of well-formulated formulas (axioms) with the aid of syntactic rules (mechanical rules of substitution), must admit propositions which are properly formulated and in a sense even true yet unprovable. Had Gödel not proven his incompleteness theorem, we might have been able to claim that even in a large formal system one need not distinguish in principle between formulas which are not well formulated and formulas which are not demonstrable, since both lead to a contradiction of a proven formula. We might have been able to argue further that one need not know in advance that a formula is not well formulated nor need one establish, in addition to the rules of inference, special syntactic criteria for the propriety of formulas, since the rules of inference suffice to show, though not at first glance, that an improper formula is self-contradictory. Roughly the argument would have looked as follows: not only the formula $2 + 2 = 5$ contradicts $2 + 2 = 4$, but also the formula $2 + 2 = +$ leads to contradicitons after a finite number of steps. If $(a=b) \rightarrow [(a+c) = (b+c)]$, and if $+(+a) \equiv +a$ then $[2+2 = +] \rightarrow [2+2+1 = +(+1)] \supset (5 = 1)$. Even a hidden lie is a lie which will always come to light. Gödel's theorem of incompleteness made any such argument impossible, since it proved the existence of well-formulated formulas which are not demonstrable, and hence also the necessity to distinguish between various modes, levels or causes of negation.

Yet precisely because such a distinction is necessary, it does not operate within one level of discourse. Not only is the one class of negative propositions not reducible to the other; they cannot be combined with each other on one level of discourse without abandoning the principle of the excluded middle. This, I argue, is the true source of the insecurity of classical logic since Aristotle, whenever it sought to distinguish between various modes of negation.

Kant had good reasons for abandoning the Aristotelian distinction between simple negations and privations in favor of another Aristotelian distinction, that between determinate and indeterminate negations.[6] 'Privation' (στέρεσις) is a state in which a subject lacks a predicate (attitude, form) which it could possess 'by nature'. In a strict sense, the term stands only for one of a pair of contrary qualities, as when I say of Socrates that he is blind. Aristotle felt uncomfortable with the negations of some privations, and rightly so, because a privative judgment both denies (that Homer can see) and affirms (that to say of Homer that he sees or that he is blind does not constitute a category mistake). What then, does a negation of such a privative judgment state? It either negates the affirmative aspect of a privation or its negative aspect, but not both. It is, of necessity, ambiguous — because it combines negations on two levels of discourse.

Moreover, even if, in the case of genuine contrarieties, the negation of a privation amounts to an affirmation (as when I say that Homer is not blind), yet if a range is involved within a quality — the Middle Ages spoke of a *latitudo formarum* — the negation of one extreme does not imply the other; not-cold does not imply hot. Such a negation constitutes an 'indeterminate negation' (ὄνομα ἀόριστον) — much as any negation of a quality which is not one of a pair of contrarieties. Now genuine privations had, for Aristotle, a distinct ontlogical status. Form and privation, as contrarieties, are 'causes', i.e. constitutive principles of any being (οὐσια), and they presuppose a third cause underlying both — namely matter — which permits a being to assume or not to assume a given form which is 'natural' to it. It is not of the nature of paper to either see or be blind.[7] In many ways, the abandonment of matter as a principle of individuation — be it with the introduction of individual forms (Scotus) or with the abandonment of the need for such a principle altogether (Ockham) — already undermined the ontic status of privations and enhanced the interest in indeterminate negations. At any rate, as from the seventeenth century the ontological meaning of privation became altogether untenable for natural philosophers, for whom nature was uniform and homogenous; they exchanged Aristotle's hierarchy of 'qualities' and 'natures' for quantitative, universal laws of one 'nature' which are applicable to all beings. To them, the 'nature' of a physical and even metaphysical subject is nothing but the sum-total of its predicates. On the other hand, the quantification of qualities became, in the seventeenth century even more than in the thirteenth, a problem of prime physical importance, e.g. in the estimation of forces and the dispute over the *vis viva*. With it also grew the interest in indeterminate (or 'infinite') judgments.

Kant removed infinite judgments from the domain of formal logic. 'Logic', for Kant as for Aristotle, was logic of terms; 'formal' logic disregards any content of the terms in a proposition — we would say: it treats all categorematic terms as variables. The infinite judgment, he claimed, belongs rather to the domain of transcendental logic, a logic which does not abstract from all content, but is concerned with all the possible contents of terms: not with any concrete content, but with the preconditions for having content. It constitutes, one might say, a first-level 'interpretation' of formal logic. As a term, the expression 'non-P' can be handled as any positive predicate: there is no need to single out, from a formal point of view, judgments containing such terms. But in view of possible content, 'S is non-P' instructs us (1) S is *not P* and (2) that S is a proper subject which belongs to the (possibly infinite) set of all subjects of which P cannot be predicated. The predicate non-P is complex: positive in its form and *limitative* in its meaning. It denies of S at least one predicate but leaves as a possibility all possible predicates. Hence it is formulated in view of all possible predicates, so that

$$[(\neg P_i) \wedge (P_1 \vee P_2 \vee \cdots \vee P_{i-1} \vee P_{i+1} \vee \cdots P_n)] \equiv \text{non-P}.$$

Again, as in the case of Aristotle's privation, a simple argument suffices to show that, against his explicit wish, Kant in fact abandons the principle of the excluded middle. The negation of an infinite negation (S is not non-P) negates either that S is not P or that S is a proper subject at all, but not both: wherefore it is not tautologically true that (not non-$P_i \equiv P_i$). But Kant was unaware of this, in spite of his observation that the infinite negation unites affirmation and negation. (This unity of two contrary forms of judgment within a third repeats itself in all four classes of judgment and their corresponding categories. The singular, infinite, disjunctive, and apodictic judgments are necessary for transcendental logic only, unite the preceding disjunction, and thus may be seen as an anticipation of Hegel's dialectical method in a precise sense.)

The infinite judgment is the paradigm and the source for a most important structuring pattern of our experimental knowledge — namely the category of limitation. This category will allow us eventually to quantify qualities, inasmuch as we regard the full presence of a quality in a subject as 'reality', its total absence as (simple) negation — zero value — and any partial presence of it, or degree, as 'limitation'. 'Quality' thus defines a continuous range and becomes quantified. The category of limitation allows us to synthesize phenomena — say the force of attraction between bodies — as *intensive magnitudes.* It is obvious that Kant here systematized the dispute over the nature of intensive magnitudes, a dispute with which he once started his academic career. Since Leibniz the dispute focused on the notion of force as *vis viva*, but, as already mentioned, had its origin in the inclination of scholastics to find a proper mathematical description for the quantification of qualities (*latitudo formarum*).[8] Herein lies also the origin of quantifying qualities with the infinite judgment. Here, as in other instances, Kant's position is a creative systematization of an ongoing dispute with a long history.

Kant moves on to prove that the categories of quality, like all others, are not only capable of structuring our experience, but that they actually do so. The category of reality can be 'mapped into' our (inner) experience of time. The act of perception in time implies a scheme of the desired category, e.g., the more intensive a perception, the more it is experienced as real; and perceptions cannot but change gradually. The 'scheme' of a category within the sense of time links elements which seemingly have no mediation: judgment and sensuality; yet without mediation there would be no cognitive ordering of experience.

The infinite judgment is thus a necessary principle for every intelligible experience — from mere sensation to the formulation of laws of nature. After many transformations from a form of judgment, through categories and then a schema, into a principle of interpretation of nature, it permits us to formulate 'synthetic *a priori*' propositions concerning acceleration and force in physics, such as Newton's first three principles. So much, then, for the role of the infinite judgment in understanding, i.e. science. Yet the same infinite judgment governs not only our understanding of experience, but also our understanding of understanding itself — the reflexive effort of reason (*Vernunft*). This effort can be critical (as when we order and legitimize our categories of understanding) or noncritical, speculative, thus leading to hypostatizations.

Kant construes a direct link from the infinite judgment and the category of limitation to the principle of complete determination. The former is transformed almost of itself into the latter. Complete determination assumes a set of all simple perfections; Kant, we recall, agreed to call them 'realities', since every quality represents a reality *sui generis* and in full presence corresponds to a sense of reality. Analogously, "reason" is led almost naturally to regard the whole range of simple predicates as the sum-total, the maximum of reality, and each predicate as one of its degrees — in the very same way in which "understanding" detects degrees *within* each quality (reality). 'Reality' is transformed, in the speculative extrapolation of reason, from a category applicable to qualities into a quality in itself. Rather than being, as it was in the effort to understand nature, a common denominator of qualities, 'reality' has become itself a quality, of which the various qualities — the simple predicates, the 'realities' — are degrees. Once the sum-total of all simple predicates is thought of as a hypostatized entity possessing the maximum of reality, all other entities can be compared to it and measured against it as various partial degrees of the same reality in comparison with 'the most real being'. Here we have the best example of how the procedures of understanding are "objectivized, then hypostatized, finally personified."[9]

Here, as elsewhere, one is struck by the architectonic precision of Kant's system. The same figures of interpretation reappear at all levels of discourse. The 'infinite judgment' appears first as a figure without interpretation (and therefore without a function) in formal logic. Transcendental logic endows it with an interpretation. In the tables of categories the infinite judgment is interpreted as a limitation, the basis for quantification of qualities. In the act of sensation, as manifested in the pure form of intuition (time), it is 'pictured' or 'schematized' in the intensity of a sensation. This leads to the objective rule of the infinite judgment in the interpretation of nature by way of the principle of intensive magnitudes ("The real element which is the object of a sensation always has an intensive magnitude, i.e. a degree"). Within the domain of pure reason, the infinite judgment permits the transformation of the concept of a 'sum-total of all possibilities' into that of 'a most perfect being' or most real being. The methodological concept of God was banned from the interpretation of nature, but it still retained a certain role as a regulative ideal of reason (as the principle of complete determination). In our language we might say: God remained, even in the *Critique of Pure Reason*, a metatheoretical assumption, an assumption which, albeit redundant in the explanation of nature, is nonetheless almost 'natural' to our reason. Kant expelled the methodical concept of God from the theory of science and grounded the universality of natural law and uniformity of nature without it; but its shadow persisted. The concept of God, he argued, is a natural shadow or projection of principles we use to structure nature. And the shadow, Kant seems to have claimed, is virtually inescapable.

Hermann Cohen's Abolition of Absolutes from the Theory of Science

Hermann Cohen exorcized from the foundation of the sciences even this shadow which remained of the methodological God; the detheologization of the sciences was to be complete. He exorcized the shadow in that he destroyed its abode — namely in that he abolished the sharp, Kantian demarcation lines between 'understanding' (*Verstand*) and 'reason' (*Vernunft*). Kant believed in a passive, receptive element within intuition; because this element can neither be isolated nor defined in itself, things-in-themselves cannot become the objects of our understanding, of our experiential knowledge. Cohen denied any *passive* residuum within cognition. Standing in this respect well within the German idealistic tradition, he insisted that all components, even objects of our cognition (or experience), have their source in the spontaneity of the mind. Yet the manner in which he abolished the dichotomy between the passive sense-data and the spontaneous structures which reasoning imposes on them, differs radically from other German idealistic endeavors to develop the manifold contents of the absolute out of its very notion. Cohen's starting point is not the absolute 'I' or the notion of an absolute, but the actual scientific achievements as a real, concrete, dynamical context embodied in concrete problems and in their solution. Within the context of science, he argues, 'facts' and the laws or theories which explain and order them are throughout interdependent. A 'fact' is not merely a 'datum' (*gegeben*) which enters the body of science from the outside; a fact is rather the sum-total of references to it in the various laws and theories, together with an unexplained and therefore problematic residue. This residue is the fact as challenge and problem: it is always *aufgegeben* rather than *gegeben*.

Cohen's main systematic exposition of his mature philosophy of science — *Die Logik der reinen Erkenntnis* — is hard to read since it points at rather than argues. It does so not because the author shuns argumentation or prefers to be ambiguous, but because of the nature of his perception of the relation between science and the reflection about it. Understanding or perception is always a cognitive act. Every component and moment of that act ought to be seen as a creative activity. Cohen therefore resembles someone who is supposed to describe a stream without reference to its banks or to any other stationary object of reference. Knowledge always means creating and shaping the object of cognition out of cognition itself. To recognize 'something' means to understand it rationally, which again means to ground this 'something', which at first is a problem for cognition, within cognition as its integral part. The solution of the 'problem' — i.e. the explication and explanation of the object with the instruments of cognition — consists in the uncovering of the 'source' of the problem, i.e. the object (*Aufgabe*), within cognition. A *pure* cognition or knowledge (*Erkenntnis*) likewise means a knowledge which purifies or purges its objects — knowledge itself, the sciences — from all noncognitive remnants. Cohen places his reasoning from the outset within a kind of hermeneutical circle which can be breached only historically and by means of historical methods: a theory of scientific knowledge — the 'logic of science' — is only possible in that we prove by acting scientifically, in that we point at the pure cognitive origin of the concepts, methods and objects of science. 'Knowledge' and 'objects of knowledge' do not define different domains but rather different directions in the activity of reasoning, as it is manifested in each concrete act of reasoning. Philosophy of science, having destroyed the autonomous status of objects, cannot view science itself as an object. It can at best *narrate* the history of science and, by so doing, make science itself point at itself and its origin. This, I believe, is the reason for the ambiguity and apodicticity in Cohen's thinking style.

Once the barrier between understanding and reason had been removed, and with it the barrier between modes and objects of cognition, the Kantian distinction between formal and transcendental logic became likewise redundant. Logic became, for Cohen, an aspect in scientific reasoning, the reflexive act through which science constantly points at itself in that it uncovers the source and origin of its objects in itself, i.e. in cognition. Logic thus occurs not in the concept but in the judgment (*Urteil*): every scientific proposition is, implicitly or explicitly, a 'judgment of origin' which creates *and* establishes the unity of the object in and through the unity of cognition. In a language which may be alluding to Maimonides' unity of knowledge, the knower and the known Cohen states: "Die Einheit des Urteils ist die Erzeugung der Einheit des Gegenstandes in der Einheit der Erkenntnis."[10]

An example is in order. Let us try to retell the history of mechanics using Cohen's terminology. To the ancients, the contraposition of rest and motion seemed to be immediately given by sense perception. This alleged 'fact' became a 'problem', the starting point of Aristotelian physics. Hence it was assumed of bodies that they have 'an inclination toward' rest in their 'proper place' (ὁίκεως τόπος), and therefore it was likewise assumed that no body can move 'by constraint' unless another body moves it for the duration of its motion: *omne quod movetur ab alio movetur.* Some difficulties of interpretation became ever more evident in this theory. How can one explain the acceleration of falling bodies? Why do projected bodies continue to move even when the force which threw them ceased to act on them — *cessante movente*? These and other difficulties, Cohen would say, were solved in the seventeenth century by the principle of inertia.

The principle of inertia, in Cohen's terminology, changed not only the theory, but also that which hitherto seemed to be a fact — the inclination of bodies toward rest (*inclinatio ad quietem*). Galileo, who was the first to employ the inertial principle, and Descartes who formulated it, recognized — again in Cohen's terminology — that the 'fact' is 'the problem', and abandoned the Aristotelian disjunction of rest and motion for the sake of the disjunction between uniform rectilinear motion, which needs no cause and of which rest is but a special case, and change of motion — acceleration or change of direction — which is always caused by force. The uniform motion of a body in a given direction is the natural state of a body 'seen in itself' (*seorsim spectatum*), i.e. in isolation from other bodies or the medium. This, of course, is a mere hypothesis, since no actual body in the universe could be far enough from the force field of other bodies so as to perform the pure inertial motion, and if there were such a body we could not observe it (for the observer would have to approach it at a finite distance and thus act upon it). The inertial principle is a hypothesis or — as we might say — a counterfactual conditional. Yet it is a counterfactual conditional with a special heuristic function: it describes the limiting case which is approached by a series of real phenomena when one or more variables (say, friction) diminish continuously.

Just as rest now became a beginning of motion, so also uniform motion became a beginning of acceleration, its 'differential'. Once the link between force and acceleration was recognized, forces could be quantified, even if only as an intrinsic magnitude (Leibniz' *vis viva, m·v²*). Such, in rough outlines, is Cohen's account of the rise of early modern mechanics, or generally of the progress of science. A 'fact' becomes a 'problem', and the contradiction between the unexplained rest of a theory (the fact) and the theory itself generates a new interpretation which explains the contradiction between concepts by making one the limiting case, the 'beginning', the 'differential' of the other. The new and better theory establishes a continuous *range* between concepts which were hitherto contradictory and hence alien to each other. Because they were alien to each other, unexplained, they could be taken to be 'given' sense-data. Then, according to the new theory, the unexplained and unquantifiable 'rest' of the Aristotelians was transformed into the 'origin' of motion, its limiting case. Correspondingly, the concept of a physical 'body' changed from 'substance' or 'underlying, undefinable matter' into 'mass-points'. Science thus creates both its facts and their explanation. One can discern, in Cohen's philosophy of science, an elaboration of Vico's principle *verum et factum convertuntur*, a guiding principle in the self-understanding of science since the seventeenth century.[11] Cohen, said Natorp, transformed the *factum* into a *fieri,* given facts into a contextual 'process' within the self-asserting reason.

Science, in other words, is a search for ever richer relations of continuities between concepts. Continuities are found whenever seemingly opposite concepts are shown to be correlative in such a way that one becomes the 'origin' or 'beginning' of the other, i.e. the constructive limiting case or 'differential', or again the ground (principle) for constructing both. The method of origin, the principle of beginning (*Ursprungsprinzip, Ursprungsdenken*), was Cohen's most central heuristic device from the very outset of his occupation with philosophy of science. It served him later in the construction of ethics and remained central even in his latest philosophy of religion, where its expression comes very close to Heidegger's (and Gadamer's) hermeneutical circle. The paradigm underlying the principle is calculus, mathematical analysis. We ought to distinguish between Cohen's material (or direct), analogical and metaphorical applications of that paradigm. Concepts which belong to a quantifiable domain in the exact sciences owe their existence to calculus. Epistemology, as we shall see, uses the paradigm of calculus as a methodological *analogy*. In ethics and philosophy of religion, it is used in a rather *metaphorical* vein.

The principle of beginning which underlies the logic of science is anchored, like Kant's category of limitation, in the infinite or privative judgment. It is safe to say that, just as in Hegel's *Science of Logic*, the qualitative categories of Kant precede the quantitative (inverting the Kantian order), so also in Cohen's *Logik des reinen Denkens*, where the category of limitation rises to the status of a monarchic principle. The genuine infinite judgment '*S* is non-*P*' is not, as Kant willed it, the paradigm of one category only (limitation); it underlies every genuine act of synthesis of concepts or images. 'Non-*P*' is both the negation and affirmation of *P* in the sense that it originates *P* without taking the values of *P*. 'Non-*P*' is the limiting case, the 'differential' of *P*. It is not as if, as the late S.H. Bergmann remarked in the most lucid exposition of the principle I know of, Cohen abandons the infinite judgment and returns to the Aristotelian usage of privation. In '*S* in non-*P*' no putative 'nature' of the subject is ever involved, such that the absence of a specific attribute which is natural to *S* could be defined as a privation.[12] Cohen rather merged the infinite and privative judgment into one mode; it serves as a methodical guideline for generating, when we wish to define the content of a concept, the closest concept from which it is distinct, and thereafter making the contradicting concept a limiting concept of the first. The very concept of 'reality' is also a case in point. It seems at first to oppose nothingness; Cohen explicates it as the infinite negation of nothingness — i.e. 'position' (*setzen*). Every act of reasoning is such an act of positing, i.e. of constituting a unity within a multiplicity and a multiplicity within a unity (or rather: within a continuous range). Every genuine act of reasoning, since it is structured by the principle of beginning, is both analytic *and* synthetic. (We remember that for Leibniz and Solomon Maimon — the thinkers closest to Cohen — such was the character of the divine mind only.) 'Beginning' has the connotations of both temporal origin and principle; the 'method of origin' is also a historical method, and Cohen, like Hegel in the *Phenomenology of the Mind*, undertook the task of proving that such indeed was the progress of scientific reasoning.

It seems to me as if Cohen allows the dialectical formation of new theories from old to come about in three steps which are repeated on ever higher levels *ad infinituum*. *P* and non-*P*, say rest and motion, appear at first in contraposition — and define each other through this contraposition. They appear as irreducible sense-data, yet their rational content lies merely in this contraposition. Once the contraposition proves insufficient to explain the contents of the contradicting elements or 'facts', the 'facts' are converted into 'problems'. Only a richer structure than a mere contradiction can link them: non-*P* is converted into a genuine non-*P*, say rest as the negation of motion into the genuine privation of motion — the instantaneous motion, the 'beginning' of motion. The reason for both *P* and non-*P* is now expressed in the generating principle of both *P* and non-*P*, of affirmation and privation, as constituting correlative elements of a continuum. It seems as if the judgments of origin envisaged by Cohen defy the principle of the excluded middle, but they do so only partially, in a carefully limited domain which forms a conceptual continuum. *Constructive* use is made of the *ambiguity* of non-*P*: it defines the range of values of *P* without taking a value itself. The new, richer concept of 'both non-*P* and *P*' is once again defined first in contraposition, then (with an even newer theory) in a continuous link to another concept. Think again of motion. Newtonian physics saw rest as the beginning of motion, but it absolutely differentiated between uniform (rectilinear) motion, which needs no external causes or forces, and change of motion, which is always due to forces. The disjunction of uniform and accelerated motion is absolute because forces are absolute, a 'fact'. In order to introduce absolute forces Newton needed absolute motions, and hence an absolute (resting) space. Another, derivated 'fact' or fortunate coincidence was the exact matching of gravitational and inertial mass. When the theory of relativity, so runs the Neo-Kantian argument of E. Cassirer, removed the absolute disjunction of inertial and accelerated frames of reference, space ceased to be absolute (in Newton's sense) and gravity (or other forces) could be interpreted as curvatures in space rather than as 'facts'. The correlation between inert and gravitational mass ceased to be a fortunate coincidence. In short, the 'residue of intuition' or 'fact' or, again, the 'sense-datum' in a given scientific theory stands for those elements within the theory which could not as yet be *derived* from it. Cohen reinstated Kant's thing-in-itself into the domain of understanding. Phenomena and things-in-themselves are two complementary aspects of the *process* of knowledge. A scientific term can be seen as reflecting all scientific propositions referring to it. It represents the sum-total of all laws relating to it; if all laws concerning (x), in the case of an ideal never reached complete knowledge of (x), were known, (x) could be named a 'thing-in-itself'. If, on the other hand, we attend to the unexplained residue of 'fact' represented by a term in the actual state of every scientific theory, (x) or the residue of the factual is the limiting case of our knowledge *of* (x). And the more our knowledge of (x) progress, the more hitherto unrelated 'facts' will be linked to it and increasingly transform the residue of fact into a concept. If our knowledge of x were complete (but it never will be), it would include all knowledge there is of other things; like Leibniz' monads, single terms would be a veritable *mundus in gutta*. The knowledge of one 'thing' would reflect the knowledge of all things; a single law would mirror all laws of nature. The complete (ideal) knowledge of a thing is the integral of knowledge, the crossing point of all laws concerning that thing, just as the 'underivable residue' of fact is, in Cohen's terminology, the 'differential' of our knowledge.

Once the thing-in-itself returned within the horizon of cognition — as an infinite task — Kant's principle of complete determination became superfluous even as a regulative ideal, and with it the methodological God even as a mere metatheoretical supposition. Instead of the metatheoretical principle of complete determination in which things-in-themselves function as abstract, mechanical limits, Cohen uses the principle of infinite negation in a different manner. He interprets the infinite negation as a dynamic 'principle or origin' both within every theory and in the reflexive assessment of science, or the conditions for constructing theories. The principle of origin is a *constructive* version of Kant's principle of complete determination, which served for *critical* purposes only. It does not 'list' ideal simple predicates but shows how one predicate (determination) generates another throughout all possible knowledge. The only ultimate guarantee which Cohen still needed to ensure the rationality of one experience (i.e., science) was the principle of generative continuity of our consciousness, which indeed ensures that we can proceed from a limit through a limit within the limit and so forth without end: *conscientia non facit saltus*. Cohen needs no absolute pattern for the contents of science, as e.g. Kant's statical 'categories' of substance, reality, causality and so forth. Cohen certainly needs no content determined by that which is outside science. The principle of beginning is not a pattern of content but a process — the process of establishing continuity between concepts by making constructive usage of their ambiguity.

But absolutes have some nasty habits. Thrown out of the main entrance, they gain reentry through the chimney. Cohen could not, after all, do without a metatheoretical substitute for Kant's methodological idea of God (or of the positive totality). Against his will he soon had to transform the principle of beginning itself into such an absolute idea. The principle became the hypostatization of the notion of continuity in physics — an embodiment of Leibniz' *lex continui*. Against his better conviction and without ever admitting it, Cohen was soon obliged to treat the principle of beginning as if it were an absolute idea, to separate the method of origin from the idea of beginning. As a methodical principle, we are taught (1) how thought begins the process of generating (*Erzeugung*) knowledge with any concrete 'problem' which it sets for itself; (2) that this process is an incessant, continuous generation of concepts from each other by way of infinite negations; (3) that this process will always leave a seemingly unexplainable residue of 'intuition' (*Anschauung*) — a new 'problem' to be subsumed under a new, unpredictable beginning. Contrary to Hegel's *Wissenschaft der Logik*, Cohen's dialectical process is infinite and open-ended. Since, however, the principle of beginning points to a process rather than a model, it cannot itself be made an object of cognition or be 'generated from' another concept. It is the *critical* task of the philosopher — and Cohen insists that the task of a philosopher is always critical only — to uncover the transcendental patterns of a given science at a given stage in its development, to lay bare its 'categories'. Unlike Kant's rigid, discrete categories, Cohen's *Urteilsformen* are but functional *a priori* structures — he calls them 'directions' — of the concrete context of scientific proportions at a given period. They generate each other (as do Hegel's categories). The principle of their generation, Cohen's dynamic equivalent to Kant's transcendental unity of the apperception, can, however, not itself be a category — let alone a concept. If it were a concept or even a concept of concepts, it would lead us straight into paradoxes of self-reference, and invite its own abrogation.[13] For if the principle of origin is a concept, we may just as well ask for *its* beginning. Does the concept of continuity between concepts emerge out of its opposite, the concept of discontinuity, and can both be linked by a different continuity? To evade such and similar paradoxes, Cohen had to separate the method from the idea of beginning. As an (absolute) idea it is just that — the never realizable idea of the unity and spontaneity of all possible knowledge. As an idea it stands outside the continuum of concept and is not a member of itself, an absolute, never exhausted paradigm, neither generated by our knowledge nor by itself. The question of its concrete origin is meaningless. In other words, if asked why science strives for unity and coherence, Cohen could only answer: because we want it so; *sic volo, sic iubeo*. Science as such, the fact of science, is a *Faktum der Vernunft*, just as freedom was to Kant a *Faktum der praktischen Vernunft*.

The End of Absolutes

The internal rift in Cohen's system — the ambiguous status of his 'principle of origin' as a method and absolute idea — could be spanned in two distinct ways. One could save the method by adandoning Cohen's rigid idea of absolute rationality, or save the idea by replacing the method.

Cassirer took the second option by systematically drawing the conclusions of the Neo-Kantian theory of science. Cohen realized very early in his career that whenever Kant spoke of 'experience' or 'nature', he actually meant the present body of scientific knowledge. Out of this hidden vice, Cohen made an explicit virtue: the moon, the used to say, does not exist out there in the heavens; it exists in our astronomy. As a consequence, his science expands like space in modern cosmology — namely in itself and not 'into' another space. It expands by inventing ever more subtle conceptual distinctions, and by linking together ever more concepts, thus creating new facts which again call for ever sharper distinctions. Science is thus a dynamic, self-sufficient context — a culture. Cohen himself paved the way toward the relativization of science: another culture may use an altogether different system of categories, of hypotheses, so as to orient itself in the world, to create its own knowledge which may be altogether incommensurable with ours. Cassirer drew this latter conclusion from Cohen's radical contextualization of science. In his *Philosophy of Symbolic Forms*, he abandoned the idea of the ultimate unity of science, an idea which already for Cohen, and even more so for Natorp, seemed a mere 'task'. Myth, so Cassirer instructs us, is neither prelogical nor paralogical, neither prerational nor irrational. It is an independent, legitimate symbolic structure governed by its synthetic capabilities. The 'concrescence' of a cultural system consists in establishing ever richer interrelations between disparate elements — Cassirer's version of Cohen's metalogical *Ursprungsprinzip*.

Of course it is a mistake to confuse relativization with error. There are those who claim against James, Vaihinger, Marx, Mannheim or Cassirer that their criterion of relativization abrogates itself because it refers to itself. This is not the case. The proposition "all propositions are relative" avoids the pitfalls of the proposition "all propositions are false," i.e. the paradox of the liar. The latter, by including itself, becomes contradictory; not so the former. The proposition that all assertions are relative may itself be relative, i.e. true under given conditions only. At any rate, Cassirer's relativism, though it still gives an edge to systems with a higher degree of reflexivity, i.e. to systems which take cognizance of others (such as our rational culture of myth), is far removed from Cohen's intentions. Cohen never tired of insisting on the supremacy of rationality over myth, of monotheism over polytheism. In the principle of origin he believed he had discovered the key to the rationality of the sciences.

Or perhaps the principle could be abandoned rather than the idea of universal rationality? I have already hinted at the direction which such an amendment ought to take Perhaps the principle of origin could be separated from its subjugation to the model of mathematical continuity? It should be conceded that with the aid of his analogy between the field of concepts and the mathematical continuum, Cohen contributed to the understanding of central features of physical thought from Galileo to Einstein. In my own interpretation of the different uses of thought-experiments since antiquity I was aided by the Neo-Kantian analogy to the continuum. Aristotle, as I have shown, already uses elaborate thought-experiments, but he does so for the sake of *reductio ad impossibile* only — for example when he almost formulates the inertial law as the absurd consequence of the absurd assumption of motion in the void. During the Middle Ages counterfactual hypothetical conditions were used not as proofs of impossibility, but for the purpose of expanding the horizon of God's omnipotence: *de potentia eius absoluta* God can create all that does not involve a logical contradiction, and can even cause our entire universe to move indefinitely and rectilinearly in empty space. Such alternative natural orders were pursued in the Middle Ages to show the contingency to every order; but each such order is still conceived as incommensurable with the factual order of our world. Only early modern physicists, notably Galileo, started using ideal counterfactual states (such as inertial motion) as limiting cases or concepts for a series of real phenomena-limiting concepts which are *constitutive for the explanation* of nature, even though they do not *describe* it.[14]

Thus, I have no intention of denying the merits of Cohen's historical perspectives. Of course, much of the concept-formation in classical physics involves the continuum. But this is not necessarily the case in science as such or in modern mathematics or physics. Already in Cohen's time, mathematicians knew of continuous functions without a derivative at any point. In the quantum, physics discovered an absolute, precise limit of material continua. All this could perhaps still be formulated using some of Cohen's terminology — provided that we give up Cohen's hypostatization of the continuum.

The principle of origin directs us to exploit conceptual ambiguities constructively. It allows us to transform conceptual contradictions into well-confined and well-defined ambiguities. Aided by this principle, Cohen formulated a generative grammar of sorts for sciences and cultures. The literal and analogous proximity of the principle to the model of continuous differentiable functions in mathematics led, so we argued, to paradoxes of self-reference. The only remedy seemed to be a strict (and, in Cohen's sense, illegitimate) separation between the principle as idea and as method. But could we not accept the paradox itself and employ it constructively? Indeed we can, provided we depart from the *lex continui* and call the infinite negation by its proper name. It is nothing but the employment of a three-valued logic within a well-defined domain. Its methodical use lies in that it shows one how to link disparate or even contradictory elements within a theory, how the ambiguity of concepts can be explicated so as to allow a precise construction.[15] Seen in this light, Cohen's logic of science serves theories of science today. Let Cohen's principle of continuity be but one device among others in the history of science. Whenever Cohen applies it within science, it refers both to the continuity it has established in this concrete case and to itself; it invited its own abrogation. Now take the principle of continuity as a beginning: Is not its negation a principle of noncontinuity, and could we not construe both as correlative principles? Perhaps, then, 'correlation' (in Cohen's sense) does not entail anything more than complementarity.

Assumptions which may be contradictory if applied to a predefined entity, an entity, so to say, identifiable by a private name, become complementary when applied to elements within a system which cannot be completely described without the system. Such is the case with the wave- and particle-nature of light; or again with the uncertainty principle, which leads to contradictions only if we assume "this electron" to be an entity independent of the system of observation. Indeed, complete description and the principle of uncertainty are not a contradiction in terms because they refer to a whole system. In Cohen's terminology we may say: the ambiguity turned into a well-defined uncertainty. His principle of beginning is a logic of explication which must always start with ambiguities, so as to remove as many of them as possible — without ever completely succeeding.

The logic behind conceptual shifts is often a dialectical one. If we choose to seek, as did Cohen, a mathematical example or analogy, we shall find it not so much in classical analysis but in the manner in which mathematical logic made constructive uses of ambiguities or even paradoxes — such as paradoxes of self-reference. The proposition "This formula is false" was transformed, in Gödel's incompleteness proofs, into a positive, noncontradictory proposition of utmost import. The key to his taming of the paradox was the manner in which the paradox was split between two levels of discourse — mathematics and a formal system large enough to formalize arithmetics. By mapping all formulas into arithmetics (with the aid of Gödel-numbers), he transformed the proposition "This formula is true only if it is false" into the proposition "This formula in arithmetics is true only if it is not demonstrable." A proper recursive formula of this sort can be construed within every formal system from whose axioms the true propositions of arithmetics can be derived, whereas arithmetics is consistent if and only if it can be proven to be incomplete. In Cohen's terminology we may say: the contradiction has become a constructive principle which calls upon us to make a new 'beginning', to decide freely whether or not a proposition proven undecidable should be made into an axiom. The history of mathematics may in a way be read as a realization of Gödel's theorem: whenever the notion of a number threatened to become contradictory and to destroy the foundations of mathematics, its meaning was expanded by further well-defined exemption from the principle of the excluded middle. In this way the Greeks discovered that a magnitude need not be *either* odd *or* even.

The scientist of today has ceased to fear the paradoxes which threatened the foundations of his discipline at the turn of the century. Not only has he learned to live with them, but also to use them and look for them. Like the Queen in *Through the Looking Glass*, he manages daily to think six impossible things before breakfast. Affinity for paradoxes and hermeneutic circles characterize many disparate areas of our culture. They have become constructive because we have abandoned the ideals of science mentioned at the start of this essay: the ideals of total uniformity of nature and transparency of the scientific language. The correspondence between them is now conceived as partial and incomplete. Absolutes were driven not only out of science, but also out of its theoretical foundations. The conclusions drawn from this state of affairs in line with some recent theories of science amount to the abandonment of the ideal of rationality. If theories, or indeed language systems in general, cannot really be translated into each other, because no agreement could be reached on synonyms (or on analytical propositions),[16] they are indeed incommensurable. Analyticity, even if it cannot be instantiated in any but thoroughly formal languages, may still, however, be an ideal. In other words, precision and transparency of terms may be regulative ideals which encourage us to look for ambiguities and remove them even at the price of other, more subtle ones. Rationality, we may hold with Cohen, does not consist of a canonical list of principles or formalizable procedures. It is a dynamical and always reflexive process which governs our theoretical enterprises without itself being capable of a clear and distinct articulation other than its instantiation.

Philosophy of science since the seventeenth century has asked time and again: What guarantees the rationality of the universe? Why is nature "written in mathematical letters"? What guarantees the compatibility of two different ideals of science — the ideal of unequivocalness of our scientific language and the ideal of uniformity of nature itself? Cohen helped us to abandon these and other ideals as absolute postulates. The question whether nature is completely symmetrical became an empirical question which can always be only partially answered. Ambiguities can never be totally removed, nor is there a universal procedure to ensure their gradual removal. *"In der Wissenschaft geht alles so herrlich und frei zu wie in einem Märchen."*[17]

Notes

1   The paradoxes of induction involve either the syntax of law-like statements (such as the Raven Paradox) or their predicates (such as Goodman's Paradox). In the latter case, the paradox applies as much to the possibility of verification as it does to the possibility of falsification. Every inductive generalization which summarizes previous observations in the forms $(x)(A(x)B(x))$ can be said to be either 'verified' or 'not falsified' irrespective of whether we shall find in the future that $B(x)$ or that $B(x)$ is the case, since it is quite possible that the property (predicate) that we interpreted as $B(x)$ in the past was, in reality, another property $B(x_{t-o})$ $B(x_{t > o})$, which behaves like $B(x)$ in the past. Cf. N. Goodman, *Facts, Fiction and Forecast* (Indianapolis, 1965), pp. 59–83.

2   G.W. Leibniz, *Philosophische Schriften*, ed. C.I. Gebhart (Berlin, 1885; reprint Hildesheim, 1965), IV, p. 6; VI, pp. 50, 321; VII, pp. 278, 303, 304, 603. Cf. A. Funkenstein, "The Dialectical Preparation of Scientific Revolutions," in: *The Copernican Achievement*, ed. R. Westman (Los Angeles, 1975), pp. 165–203, esp. 178–187.

3   William of Ockham, *Centiloquium theologicum,* concl. VI. Both in my article (note 2) and elsewhere ("Continuity and Change in 17th Century Thought and Science," *Publications of the Israel Academy of Sciences*, VI, 6, Jerusalem, 1981, pp. 101–131) I have argued that, in contrast to the Middle Ages, science in the seventeenth century employed counterfactual states — ideal experiments — not to demonstrate God's omnipotence, but as limiting cases of reality. Such thought-experiments (e.g. inertial motion) do not describe reality, yet are constitutive for its understanding. Cf. p. 59.

4   Thomas Aquinas, *Summa Theologiae* 9 q. 25 a. 5 resp. 3; *De Potentia* q. 3.a. 17, p. 103. In both

places, Aquinas employs the remarks of Maimonides about the limit of every rationalization — that is, the element of contingency necessary in every conception of universal order. Cf. my comment in *Miscellanea Medievalia* XI (1977), p. 89, n. 27.

5 Immanuel Kant, *Kritik der reinen Vernunft* (hereafter *KdrV*), *Gesammelte Schriften* (Berlin, 1911), B 599–611.

6 *Ibid.*, B 95–98.

7 Aristotle, *Metaphysica* V, 1022b22.

8 A. Maier, *Kants Qualitätskategorien* (Kantstudien 65), Berlin 1930, pp. 8–23; "Die Mathematik der Formalattituden," in: *An der Grenze von Scholastik und Naturwissenschaft* (Rome, 1922); E. Sylla, "Medieval Conceptions of the Latitude of Forms — the Oxford Calculatores," *Archive d'Histoire doctrinale et littéraire du Moyen Age* 51 (1974): 223–283.

9 Kant, *KdrV*, B 611 (note).

10 Hermann Cohen, *Logik der reinen Erkenntnis* (Berlin, 1914²), p. 68. In the following, I have not attempted to root out the shift from Cohen's earlier perspective (in: *Das Prinzip der Infinitissemalmethode und seine Geschichte* (reprint, Frankfurt am Main, 1968)) to his mature theory of science.

11 B. Croce, *Die Philosophie Giambattista Vico's*, transl. E. Auerbach and Th. Lucke (Tübingen, 1927), pp. 3–17; K. Löwith, "Vico's Grundsatz: verum et factum convertuntur," in: *Aufsätze und Vorträge 1930–1970* (Stuttgart, 1971), pp. 157–188.

12 S.H. Bergmann, "The Principle of Apriority in the Philosophy of Hermann Cohen," *Knesset* 1944; *Introduction to Logic* (Jerusalem, 1953), pp. 210–213. Cf. also W. Marx, *Transzendentale Logik als Wissenschaftstheorie* (Frankfurt am Main, 1977), pp. 103 ff., esp. 119 ff.

13 Since this is so, W. Marx argues that Cohen never succeeds in establishing science as the exclusive property of pure thought (*op. cit.*, pp. 133–154). But he does not treat self-referential structures in Cohen's interpretation explicitly.

14 Above note 3.

15 D. Henrich, *Hegel im Kontext* (Frankfurt am Main, 1971), pp. 95 ff., showed how the dialectical moves in the *Wesenslogik* can be interpreted as a logic of explication. Cohen, whose logic is both open-ended and nothing but a logic of scientific discovery, renders himself even better for such a perspective.

16 W.O. Quine, "The Two Dogmas of Empiricism," *From a Logical Point of View* (Cambridge, Mass., 1953); H.P. Grice and F.P. Strawson, "On Defense of a Dogma," *Philosophical Review* 65 (1956): 141–158.

17 Robert Musil, *Der Mann ohne Eigenschaften* (Hamburg, 1952), p. 40.

Origin and Spontaneity
A Comment

NATHAN ROTENSTREICH

I

As a preface to some comments on the philosophical position of the Marburg School, with special emphasis on Hermann Cohen — following the analysis presented by Professor Funkenstein — two topics of an historical and typological character should be briefly discussed.

The first topic is the attitude to the philosophical trend which is represented by Solomon Maimon. Maimon described himself as a Leibnizian who could not help accepting Hume's position. This aphoristic résumé is meant to indicate that essentially the philosophical position is destined to present the unity of sensuality and rationality. Since we, as finite beings, cannot establish that unity personified by the reference to Leibniz, the only alternative is the pre-Kantian position personified by Hume. Hume presents the duality of sensuality and rationality but questions the applicability of rationality to sensuality. Hence Maimon questions the very trend of the Kantian solution for the 'dogmatic slumber'. Philosophy is aware of the need for a solution but cannot achieve it.

Cohen, as well as the Marburg School with all its variations, do not follow Maimon's position on this issue. Though, as we shall see, rationality is not viewed by Cohen as applicable to data but as creating the objects, for the sake

E. Ullmann-Margalit (ed.), The Kaleidoscope of Science, 65–73.
© *1986 by D. Reidel Publishing Company.*

of rationality, he does not attempt to deny altogether the position of data. Data are mobile, as it were; Cohen refers to the claim of the sensation.[1] This only means that the categories would take that claim upon themselves. It would be a prejudicial assumption to say that categories are involved only in the immanent game of pure thinking. Sensation has to be taken into account as well, and consciousness would not go beyond pure thinking if it did not integrate the component of sensation as a moment for critical evaluation (*Abschätzung*). Hence Maimon's system, though Cohen does not specifically refer to it, is to be rejected — and a different trend, stressing the dynamic character of knowledge, brought forward. That trend does not recognize the rigid dichotomy between rationality and sensuality. It takes into account the eventual and perpetual integration of sensation into the dynamic structure, i.e. into the process of knowledge. We shall presently come back to that point.

II

The allusions to the process of knowledge and, as it were, to the changing positions of the datum and of *ratio*, lead us to consider whether Cohen's systematic position should somehow have been interpreted in a Hegelian manner. This cannot be the case because processuality as such cannot exhaust the respective systematic positions. For Hegel, the datum is already objectively imbued with rationality; essentially it is a given rationality, yet still lacking self-consciousness. This is one of the meanings of the renowned saying that all that is immediate is already mediated. The process is the self-exposition of the datum, and the process of knowledge cannot be considered as creation. The process makes rationality explicit, but not sovereign in the way Cohen saw it, when at this point he followed the main trend of Kant's system, viz. that rationality has a legislative quality or capacity. Thus reason imposes its structure on the data, although Cohen, to be sure, took a further step in the direction envisaged by Kant. Hence, he put it as a geometrical metaphor: knowledge starts with rationality and not with data. But starting with rationality, knowledge exhibits its sovereignty, though not its omnipresence. It is with this reservation that Cohen refers to the cold comfort (*Trost*) of the dialectical movement of concepts. The certainty of knowledge is being confuted. The central position attributed by Hegel to the principle of contradiction shows up the "abyss" of his logic.[2]

Since Hegel, as it were, starts from below and moves in an upward direction from data to rationality, he assumes that eventually a meeting of the beginning and the end is to be accomplished. This leads him to view totality as the highest order — a kind of Spinozistic substance rendered in an articulate way. For Cohen the emphasis placed on totality is considered as a tendentious mistake in Hegel's logic. Hegel is not concerned with the totality of conditions — and it is obvious that the employment of the term 'condition' has a more than Kantian innuendo. Hegel moves toward the totality of that which is (*des Seienden*). Thereby, in Cohen's view, he commits an ontological-categorical mistake.[3]

Having made these comments, we may now attempt to interpret a few of the positive aspects of Cohen's system.

The very description of reason as a faculty — as conceived by Kant — may provide it with a static capacity. It is a static capacity endowed with spontaneity or self-activity, as Kant described it. Perhaps we may say that Cohen's point of departure lies in what he considered — and this is my interpretation only — as a lack of consistency when spontaneity is understood to be grounded in a stable factor qua faculty. Hence Cohen steps "above" Kant by viewing spontaneity as such as the origin of knowledge, an origin which never exhausts itself and, by the same token, continually manifests itself in the process of knowledge.

Let us revert to a Platonic notion or term related to our topic: We have to distinguish between *Hypothese* and *Hypothesis*, since *Hypothese* implies the infinite regression from a thesis to another *Hypothese*. *Hypothesis* is the foundation which as such is not involved in the infinite regression. Its foundation or grounding (*Grundlage*) comes first (*das Erste*) in all thinking. All thinking of an idea is thinking of an Hypothesis.

The Idea is the *Hypothesis* and the *Hypothesis* is the Idea.[4] There is no need to emphasize here the change in terminology as compared with the terminology of Kant, in order to emphasize the change in orientation. Cohen does not employ the term 'Idea' in Kant's sense as pointing to the totality of reality (or to the Ideal as God). He presents the idea as the *Hypothesis*, that is to say, as the foundation of knowledge, and as such as its dynamic anchor. Since we refer to *Hypothesis*, we may emphasize that Cohen himself says that the question is that of the beginning and value of something original (*ursprünglich*), which can only be a principle (and is bound to be one). The concept is the problem of positive creation, which can succeed only in terms of deduction and never in terms of induction.[5] All these remarks lead us to comment on the notion of origin — *Ursprung*.

III

Now it is our task to explore only the *Leitmotif* of Cohen's system in terms of his adherence to the Kantian tradition,[6] the sort of adherence which became the main concern of the Marburg School. Even in terms of the architectonic structure of the system it is essential to notice that Cohen, unlike, for example, Hegel, presents the various domains of philosophic analysis as coexisting and not as forming a line of continuity or of ascent. Not only the books dedicated to the exploration of Kant's system are organized in a way parallel to Kant's three Critiques, that is to say: knowledge, ethics, and aesthetics. Cohen's own systematic books follow the same structure of coexistence in terms of pure knowledge, pure will, and pure feeling. The emphasis on the adjective "pure" is obviously intentional and points to Cohen's step beyond Kant, or, if we are permitted to use that expression, "above" Kant. We shall deal with that ιopic presently. To be sure, Cohen is more concerned with the position of psychology than Kant was: in the course of his development, he presents a new attempt in the direction of a philosophy of religion, stressing the ethical content of religion. But, unlike Kant, he presents in his opus on religion and its origins in Judaism as an ethic of the individual person and of his fellow-men, but does not identify the ethical system with the ethics of humanity, as Kant did and as Kant maintained in his book on religion. To be sure, these are significant attempts to amend Kant's presentation, but as such they cannot be understood as superseding Kant.

This direction comes to prominence in Cohen's notions of the interrelation between *Hypothesis* and origin (*Ursprung*). The following can be said by way of a résumé which precedes some details that have to be mentioned later on. Kant understood reason as a faculty of principles. Reason as understanding, that is to say, understanding as the partial manifestation of reason, manifests its priority as legislation within the spectrum of knowledge. Hence the categories legislate that which appears within the scope of experience, that is to say, that which is given by sensuality in its a priori forms and, by the same token, delineates that which is outside or beyond experience and thus beyond knowledge. Legislation cannot be valid unless it is applied to something, i.e. to data which are subsumed under the principles of legislation. Hence the problem of application of the principles to data preoccupied Kant's analysis — he reverted to it time and again. One of the major manifestations of the preoccupation is, obviously, the position of schematism in Kant's theory. For Kant the problem of schematism arises because there are two trees of knowledge: sensuality and reason (*Vernunft*), and hence understanding (*Verstand*). They might be united, but their unity is not within the scope of our knowledge. Hence, Kant looked for possible ways of establishing a harmony between the levels of forms and found harmony on the level of sensuality in the form of time. It is no accident that, when Cohen uses the term or the concept of *Schema*, he refers to the number (*Zahl*) as the principle of the extensive magnitude[7] and not to time, since time is turned into a category within the process of succession of categories. The notion of the form of sensuality or pure sensuality is imbued with a basic deficiency in that it offers only a provisional content. It is meant to contain the proper force of the content but does not fulfill that expectation.[8] It is clear that that "disappointment" is related to the philosophic position where the constant, though amorphous, status of the datum is no longer a constant position or datum.

In an attempt to present the essence of *Ursprung* the difficulty lies perhaps in its very subject matter, if the use of that term is at all permitted. Speaking about spontaneity as such, or self-activity as such, we face the difficulty of any further description, since any description may amount to a difficulty or dialectic that *omnis determinatio est negatio*. This would imply a limitation of the spontaneity which, as the origin, is meant to be unlimited. Yet we shall follow Cohen's own hints as to the essence of the origin, which brings the aspect of infinity into the horizon of philosophic analysis.[9] Again it has the general connotation of the principle and of laying the foundation, and here we refer again to the notion of *Hypothesis*.[10] Thinking is thinking of the origin, and the latter can never be determined. Only as thinking of the origin does pure thinking become truthful (*wahrhaft*).[11] Origin and creation (*Erzeugen*) are interrelated; hence Logic is Logic of the Origin. The creation itself is the created (*Erzeugen — Erzeugnis*).[12] The infinitesimal served here as a model, or in the Kantian sense as a monogram, for the process and procedure of knowing and thinking.[13] Going beyond Kant, we can say that there is a continuity from thinking to knowing. Logic itself becomes involved in the critique — again Cohen differs here from Kant — since Logic connotes to us the Logic of the Origin. The Logic of the Origin bestows competence on pure thinking.

IV

It is only consistent that Cohen related the notion of Origin also to ethics, stating that therein lies the source and the foundation of ethics, since it is, or is related to, the principle of freedom.[14] Kant saw in freedom the full manifestation of reason even more than reason as it is manifest in the orbit of knowledge. Freedom for Kant is the complete exhibition of reason as the faculty of legislation. In a sense Cohen introduced the aspect of freedom into Origin, in that spontaneity is comprehensively exhibited in the realm of knowledge as well as in the realm of ethics. Since the comprehensive Origin underlies both knowledge and ethics, we can say that the primacy of theoretical reason replaces Kant's concept of the primacy of practical reason. Yet we could rather say that *Ursprung* is above the distinction between knowledge and ethics.

We can suggest a further aspect of the essence of *Ursprung*. Since it amounts to spontaneity, there is a more basic affinity between Origin and the position of the question than between the Origin and the position of the proposition. The logical meaning and significance of the question become prominent in its position as a "lever" (*Hebel*) of the Origin.[15] In addition, since the full spontaneity is presupposed or perhaps identified, the continuity of knowledge is an ongoing process and procedure, stemming from the inexhaustible Origin. Because of this continuity, all elements of thinking, insofar as they constitute knowledge, are, or have to be, created out of Origin. In this sense the Origin is a kind of shield against the given (or the datum).[16]

To sum up, employing a terminology we do not find in Cohen's own presentation, we could say that, in terms of the Origin, no distinction is to be made between its essence and its function. The very ultimate fact that knowledge proceeds infinitely indicates the fullness of the Origin as well as its spontaneity (not to be propositionally stated). Cohen did not attempt to introduce any intuition (*Anschauung*), not even an intellectual one, into the orbit of his exploration, since he probably could not assume that reason as Origin can be identified by anything but reason itself. The problems which Cohen's system faces here are obvious: the attempt to remove the datum from the scope of knowledge, and concurrently the attempt to remove from it the aspect of application, would sacrifice the fullness — or concreteness — of knowledge for the sake of its creativity. These are critical issues which we mention here without analyzing them in detail. We shall simply wind up with a comment on Cassirer's philosophy of symbolic forms, in order to indicate a kind of internal change that occurred within the Marburg School, a change which possibly is not unrelated to the problem which we just indicated as emerging out of Cohen's system.

V

The justification for selecting Cassirer as the major representative of the Marburg School, in winding up our short exploration of Cohen's philosophy of science, is twofold: Cassirer dealt with science within the broad contours of his system mainly in terms of his philosophy of symbolic forms. The second justification relates to the fact that Cohen dealt with scientific theories in a more defined sense, namely with the problem of the foundations of mathematics, with the shift from the category of substance to the concept of function, with the theory of relativity and with causality. Eventually, he also dealt with some aspects of the humanities qua *Geisteswissenschaften*. To be sure, his historical opus has also to be mentioned in this context.

In the following we shall concern ourselves only with one aspect of Cassirer's interpretation, namely, with his attempt to reconsider the position of science within the spectrum of interpretation of the 'world'. At this point the philosophy of symbolic forms deviated from the philosophy of *Ursprung*. In this context we have to mention that neither ethics nor aesthetics were Cassirer's major concern. He remained within the boundaries of 'culture'. But within those boundaries he was concerned with the theoretical or symbolic interpretation, or productiveness, with the theory of symbols in general, and more specifically with language, myth and science. He himself said that the critique of reason became the critique of culture. Cassirer accepted the basic position of Kant — or perhaps of Cohen — that what is taken as 'the world' can never be grasped immediately; it is a productive interpretation as an element of symbolic forms. The various modes of intellectual — or spiritual — activity in the sense of *geistig* molding (forming) are not placed in one line or series — and on this issue Cassirer's system differs basically from Cohen's.[17] Yet the manifold of products does not gainsay the unity of producing; on the contrary; it reinforces and confirms it.[18]

Cassirer comes back to Kant's basic notion, underscoring perhaps more the mutuality or correlation between the sensuous and the intellectual elements than the dichotomy which has to be bridged — as Kant had it. It ought to be observed that in terms of scientific knowledge Cassirer describes the direction of his attempt as a phenomenology of knowledge and essentially reverts to Kant's distinctions. Hence, the two major parts of his volume concerned with science present the problem of representation and the structure of the intuited world (*anschauende Welt*), and the function of meaning (*Bedeutungsfunktion*), as well as the structure of scientific knowledge. This multidimensionality is therefore not only confined to the coexistence of different symbolic forms, but deals with the structure in the immanent sense of the individual symbolic forms. Philosophy as philosophy of symbolic forms therefore goes beyond the affinity between science and philosophy of foundations in the direction of philosophy or to the theory of structures of different spheres of productivity.

Having said this, we have to make a reservation about that very presentation. This reservation emerges rather from the inner logic of

Cassirer's systems than from his explicit statements. In the first place we may perhaps point to the affinity between science and philosophy, in spite of what we said previously about the multidimensionality of the symbolic forms. Science is accompanied by self-reflection, and in this sense there still exists a kinship between science and philosophy, which is stronger than that between the various symbolic forms and philosophy. The former are produced without being accompanied by self-reflection as to their own inherent structure. The aspect of self-reflection is not only a structural element inherent in science. It finds its expression in the extension of the scientific or scholarly concern: it is not philosophy proper that explores the symbolic structure of each of the elements of the 'cultural consciousness'. The scientific method or the scholarly one — the terminology does not matter in this context — refers to language and myth, at least in terms of their diffusion and dispersion, if not in terms of the discernment of their basic elements. This is a one-way street: we do not explore science from the angle of myth, but we explore myth from the angle of science, employing scientific tools duly interpreted and applied.

In any case, it might be proper to revert to Hermann Cohen and quote the well-known statement which is characteristic of his own systematic position but which, duly amended, is characteristic of Cassirer's position as well: "Not in the skies are stars *given*, but in the science of *astronomy*. We describe those objects as given. Not in the eyes lies the sensuality, but in the *raisons de l'astronomie*."[19]

Notes

1   Hermann Cohen, *Logik der reinen Erkenntnis* (Hildesheim/New York: Georg Olms Verlag, 1977), p. 436.
2   *Ibid.*, pp. 11 ff.
3   *Ibid.*, p. 329.
4   Hermann Cohen, *Der Begriff der Religion im System der Philosophie* (Giessen: Alfred Töpelmann), pp. 28–29.
5   *Ibid.*, p. 5.
6   Paul Natorp, *Hermann Cohens philosophische Leistung* (Reuther & Reichard, 1918), sums up Cohen's attitude to Kant.
7   Hermann Cohen, *Logik der reinen Erkenntnis* (see note 1), p. 480.
8   *Ibid.*, p. 150.
9   *Ibid.*, p. 35.
10   *Ibid.*, p. 36.
11   *Ibid.*
12   *Ibid.*, p. 53.
13   *Ibid.*, p. 35.
14   *Ibid.*, p. 44.

15   *Ibid.,* p. 85.
16   *Ibid.,* pp. 91–92.
17   *Philosophie der Symbolischen Formen,* Erster Teil (Berlin: Bruno Cassirer, 1923), p. 29.
18   *Ibid.,* p. 51. Relevant observations as to the difference between Cassirer and Cohen are to be found in Wolfgang Marx "Cassirers Symboltheorie als Entwicklung und Kritik der Neukantischen Grundlagen einer Theorie des Denkens und Erkennens," 2. Teil, in: *Archiv für Geschichte der Philosophie* (1975), Heft 3, pp. 304 ff.
19   Hermann Cohen, *Das Princip der Infinitesimal-Methode und seine Geschichte, Ein Kapitel zur Grundlegung der Erkenntniskritik* (Berlin: Frd. Dummler, 1883), p. 127.

13. ...

18. ...

19. ...

Cognitive Illusions in Judgment and Choice

Amos Tversky

Human errors provide valuable data for the study of the structure and the functioning of the human mind. Much research on memory is based on forgetting, work in perception is often based on discrimination failure, and the study of language and thought has exploited slips of the tongue. The study of human error is particularly important because (i) it demonstrates some limitations of the human mind, (ii) it sometimes reveals the rules or mechanisms that underlie people's behavior and (iii) it often suggests ways in which human performance can be improved.

Illusion is a special kind of error that is particularly disconcerting because the error that one has made remains attractive, even when one has been throughly convinced that it is an error. For example, the impression that the right line is longer than the left line in Figure 1, persists even after we have verified that the two lines are in fact equal in length. Besides the natural fascination with illusions, they offer the psychologists a rich source of data about human perception and judgment. For example, the Muller–Lyer illusion displayed in Figure 1 is caused, in part, by the use of perspective cues as illustrated in Gregory's analysis. The point of this digression is not to discuss theories of visual perception, but rather to illustrate the manner in which illusions provide valuable evidence about the underlying psychological processes (Kahneman and Tversky, 1982).

75

E. Ullmann-Margalit (ed.), The Kaleidoscope of Science, 75–87.
© *1986 by D. Reidel Publishing Company.*

Figure 1. The Muller–Lyer illusion embedded in a familiar scene,
after Gregory (1970)

In this lecture, I would like to describe two classes of cognitive illusion: the
conjunction fallacy in probability judgments caused by representativeness,
and the illusions of framing in decision-making that are caused by shifts of
reference points. I will present several illustrations of each phenomenon and
discuss their psychological significance and their real-world application.
This research has been done in collaboration with Daniel Kahneman.

Probability, Representativeness and the Conjunction Fallacy

Uncertainty is an essential element of the human condition. The decisions we
make, the conclusions we reach, and the explanations we offer are usually
based on beliefs concerning the probability of uncertain events, such as the
results of an experiment, the outcome of a surgical operation or the future

value of an investment. In general we do not have at our disposal adequate formal models by which the probabilities of such events can be calculated. Instead, people appear to rely on a limited number of informal rules, or heuristics, in an attempt to provide simple answers to complicated questions. In general, these heuristics are quite useful and effective, but often they lead to severe and systematic errors.

A great deal of research conducted during the last decade has shown that people often evaluate the probability of an uncertain event or the frequency of a class by the representativeness heuristic (Kahneman et al., 1982). According to this heuristic, an event is judged to be probable if it is representative of the class from which it is sampled. In other words, people often evaluate the probability that event A originates from process B by the degree to which A is representative of B. This heuristic has some validity because, other things being equal, representative samples are more probable than nonrepresentative samples and typical instances are more frequent than atypical ones. An uncritical reliance on representativeness, however, could lead to serious errors because representativeness has a logic of its own which does not coincide with the laws of probability.

An event can be improbable either because it is nonrepresentative or because it is highly specific. The probaility that the person who sits next to you in the bus will weigh more than 100 kilograms is fairly small because this event is atypical. On the other hand, the probability that the person who sits next to you in the bus will weigh exactly 70.2675 kilograms is extremely small because the event is very specific. Indeed the latter is a much more representative weight for an adult male, although the former is much more probable. As this example illustrates, an increase in specificity does not necessarily lead to diminished representativeness. A random sample of four cards consisting of the king of hearts, the ace of spades, the nine of diamonds and the four of clubs, appears more representative than a sample consisting of four cards of the same suit, although the latter is far more probable. We suggest that people tend to judge probability primarily by representativeness and neglect or underweight the specificity of the events.

The sharpest contrast between the logic of probability and the representativeness heuristic arises in the evaluation of conjunctions or compound outcomes. Suppose that we are given information about some individual (e.g. a personality sketch) and that we speculate about various attributes or combinations of attributes that this individual may possess, such as occupation, avocation or political affinity. Perhaps the most basic law of probability is that specificity reduces probability. Thus the probability that a given person is both a Republican and an artist must be smaller than the probability that the person is an artist. This condition holds not only in the standard probability calculus but also in nonstandard models (e.g. Shafer, 1976; Zadeh, 1978). However, the requirement that P(A & B) < P(B), which may be called the conjunction rule, does not apply to similarity or representativeness. A blue square, for example, can be more similar to a blue circle than to a circle, and an individual may resemble our image of a Republican artist more than our image of a Republican. This is so because the similarity of an object to a target can be increased by adding to the object features that are shared by the target (Tversky, 1977).

Similarity or representativeness, therefore, can be increased by further specification. If probability judgments are mediated by representativeness or similarity, we would expect to find situations where a conjunction of outcomes appears more representative and hence is judged more probable than one of its components. I will refer to this phenomenon as the conjunction fallacy. To illustrate this effect, consider the following description, which has been presented to many groups of subjects.

> Bill is 34 years old. He is intelligent, but unimaginative, compulsive, and generally lifeless. In school, he was strong in mathematics, but weak in social studies and humanities.
>
> Please rank order the following statements by their probability, using 1 for the *most* probable and 8 for the *least* probable.
> (3.7)   Bill is a physician who plays poker for a hobby.
> (3.9)   Bill is an architect.
> (1.1)   Bill is an accountant.   (A)
> (6.2)   Bill plays jazz for a hobby.   (J)
> (6.6)   Bill surfs for a hobby.
> (5.7)   Bill is a reporter.
> (3.4)   Bill is an accountant who plays jazz for a hobby.   (A & J)
> (6.1)   Bill climbs mountains for a hobby.

As the reader has probably guessed, the description of Bill was constructed to be representative of an accountant (A) and unrepresentative of a person who plays jazz for a hobby (J). Indeed, the overwhelming majority of subjects who were asked to rank these descriptions in terms of representativeness conform to this prediction. Furthermore, the conjunctive outcome (A & J) was judged intermediate in representativeness. This pattern is neither surprising nor objectionable. Since an accountant is more compatible with Bill's personality than a jazz player, the conjunction of the two may be expected to be less representative than an accountant but more representative than a jazz player.

What is perhaps more surprising and less defensible is the finding that the probability ordering of these descriptions practically coincided with the representativeness order. Different groups of subjects who were asked to rank order the statement by probability judged the conjunctive statement (A & J) to be more probable than its less probable component (J). (The numbers in parentheses are the mean ranks assigned to the various outcomes.)

We have observed the conjunction fallacy in a between-subjects design, in which one does not compare the critical event directly, as well as in a within-subject design involving only two descriptions. Furthermore, even highly sophisticated subjects, such as graduate students in statistics and decision science, exhibited the conjunction fallacy, although the effect there was slightly weaker. However, almost all subjects committed the conjunction fallacy in a between-subjects experiment, and most of them exhibited the fallacy even in a within-subject experiment.

We took great pains to try to eliminate the conjunction fallacy but our efforts were, by and large, unsuccessful. We have presented subjects with arguments for and against the conjunction fallacy and asked them to judge which argument is more nearly correct. In the above problem, for example, subjects are presented with the following two statements.

> 1. Bill is more likely to play jazz for a hobby than to be an accountant who plays jazz for a hobby because every member of the latter class is necessarily a member of the former class.
> 2. Bill is more likely to be an accountant who plays jazz for a hobby than a person who plays jazz for a hobby because he is more representative of the former than of the latter class.

Much to our surprise about 50% of naive subjects selected the incorrect argument. Evidently, the conjunction law, which appears simple and appealing in the abstract, is often overruled when it conflicts with a strong intuition of representativeness. Statistically sophisticated subjects, however, overwhelmingly chose the correct argument. Although the latter result is encouraging, it is generally disappointing that statistical sophistication and extensive training do not reduce the incidence of error but merely the tendency to justify it after the fact.

This set of experiments is open to several alternative interpretations in terms of the meaning of probability, the natural interpretation of conjunction, the role of conversational implicatures, and so on. To explore these possibilities we have run several additional studies, which were designed to eliminate or reduce the effect of these variables. In particular, we have observed the conjunction fallacy in problems referring to the frequency of classes rather than to the probability of unique events. For example, the median estimate of the percentage of people in a sample that are both above 35 years old and suffer from arthritis was larger than the estimated percentage of people who suffer from arthritis. Similarly, the median estimate of the proportion of people who are rock fans and under 30 years of age was higher than the estimated percentage of rock fans. Although these questions refer specifically to the frequency of classes, people evidently thought about them in terms of cultural prototypes and not in frequentistic terms. It is important to note that the conjunction effect is not due to averaging. When the two components appear independent or inconsistent, the judged probability of the conjunction is less than that of the individual components. For example, the median estimate of the proportion of people who are both older than 40 and rock fans was considerably smaller than the median estimate of each of the two categories.

The conjunction effect occurs with real-world examples and not merely with hypothetical ones. For example, we asked subjects to imagine that Connors and McEnroe are the two finalists at Wimbledon in 1982 and asked them to rank order several outcomes from most to least likely. A great majority of subjects ranked the outcome that McEnroe would lose the first set as less probable than the outcome that McEnroe would lose the first set but win the game, in flagrant violation of the conjunction rule. Similarly, most subjects ranked the possibility that Reagan will provide federal support for unwed mothers as less probable than the possibility that Reagan will provide federal support for unwed mothers and cut federal support for local governments. We have also obtained violations of the conjunction rule in jury type problems where subjects were asked to order several scenarios in terms of their plausibility. For example, subjects judged the event that the defendant committed two crimes as more probable than the event that the defendant committed one crime. Perhaps even more alarming is the finding that physicians were highly prone to the conjunction fallacy in judgments about medical diagnosis. For instance, the majority of physicians who were presented with the description of a twenty-year-old woman who suffered from infectious mononucleosis judged the probability of sore throat and conjunctivitis to be greater than the probability of conjunctivitis. Naturally, sore throat is more representative of mononucleosis than conjunctivitis, but both symptoms can never be more probable than one of them alone.

Finally, we have obtained the conjunction fallacy in a gambling task involving real money, in which no reference to probability or frequency was made. The subject was presented with a regular six-sided die, four sides of which were painted green and two sides were painted red. Each subject was told that the die would be rolled twenty times and that he could win a prize of

$25 if the sequence he selected is observed on successive rolls of the die. People were then asked to choose between two sequences: RGRRR and GRGRRR. Clearly, the second sequence is more representative since it includes two greens and four reds, whereas the first sequence includes only a single green. However, the first sequence is necessarily more probable since it is included in the second sequence. The subjects, however, apparently did not detect this relation and over two-thirds selected the second (less probable) sequence. Hence, the conjunction fallacy is manifest in decisions involving real payoffs, not only in judgments of probability or frequency.

What are the implications of the conjunction fallacy? Firstly, it offers the clearest and perhaps the most alarming demonstration of the discrepancy between the principles that govern the intuitive evaluation of events and the rules of logic and probability. Secondly, it suggests that the discrepancy between the intuitive and the formal conceptions of probability is deeper than we had originally thought. Previous work on probability judgments has led to the conclusion that the reliance on judgmental heuristics sometimes leads to predictable biases. The present results, in contrast, suggest that intuitive judgments of probability are not merely biased, they are fundamentally different from the normative model. Thirdly, the conjunction fallacy could have serious practical implications. In everyday life we are often called upon to compare the likelihood of events that differ in both specificity and typicality. The tendency to evaluate probability by representativeness and to ignore or underweight the specificity of the target event could lead to serious errors in judgment and decision (Tversky and Kahneman, 1983).

It is important to emphasize that the conjunction fallacy is merely a symptom of a more fundamental problem, namely the tendency to underweight the specificity of events in assessing their likelihood. This tendency is likely to be particularly pronounced in an attempt to forecast the future by generating various scenarios. As they stare into the crystal ball, politicians, futurologists and lay persons alike seek an image of the future that best represents their model of the dynamics of the present. This search leads to the construction of detailed scenarios, which are internally coherent and highly representative of our model of the world. Such scenarios often appear more likely than less detailed forecasts, which are in fact much more probable. As the amount of detail in a scenario increases, its probability can only decrease steadily, but its representativeness and hence its apparent likelihood may increase. A reliance on representativeness, we believe, is a primary reason for the unwarranted appeal of detailed scenarios and the illusory sense of insight that such constructions often provide.

Decision Frames and Illusions of Choice

Let me turn now to describe another family of cognitive illusions that are concerned with preferences rather than with judgment. A decision problem is normally characterized in terms of the acts among which one chooses, the possible outcomes or consequences of these acts, and the contingencies or conditional probabilities that relate outcomes to acts. It is often possible to frame a given decision problem in more than one way. The frame that the decision-maker adopts is controlled partly by the formulation of the problem and partly by the norms, habits and personal characteristics of the decision-maker (Tversky and Kahneman, 1981).

Alternative frames for a decision problem may be compared to alternative perspectives on a visual scene. Veridical perception requires that the perceived relative height of two neighboring mountains, say, should not reverse with changes of the vantage point. Similarly, rational choice requires that the preference between options should not reverse with changes of frame. Because of imperfections of human perception and decisions, however, changes of perspective often reverse the relative apparent size of objects and the relative desirability of options. We call the reversal of preferences that is induced by altering the description of outcomes a framing effect. Framing effects arise when the same objective alternatives are evaluated in relation to different points of reference. For example, we asked several groups of physicians to consider the following problem.

> Imagine that the United States is preparing for the outbreak of an unusual Asian disease which is expected to kill 600 people. Two alternative programs to combat the disease have been proposed. Assume that an exact scientific estimates of the consequences of the program are as follows:
> If Program A is adopted 200 people will be saved.
> If Program B is adopted there is a 1/3 probability that 600 people will be saved and a 2/3 probability that no people will be saved.
> Which of the two programs would you favor?

The majority response to this problem is risk-averse preference for Program A over Program B. Other respondents received the same cover story with a different formulation of the alternative programs as follows:

> If Program C is adopted 400 people will die.
> If Program D is adopted there is a 1/3 probability that nobody will die and a 2/3 probability that 600 people will die.

The majority choice in this problem is risk-seeking: the certain deaths of 400 peope are less acceptable than the 2/3 chance that 600 people will die. It is easy to see, however, that the two versions of the problem describe identical outcomes. The only difference is that in the first version the deaths of 600 people is the normal reference point and the outcomes of the programs are evaluated as gains (lives saved), whereas in the second version, no deaths is the normal reference point and the programs are evaluated in terms of lives lost. The shift of reference point is accompanied by a dramatic shift from risk-averse to risk-seeking preferences. This is a prevalent and robust finding observed in many groups of respondents using hypothetical as well as real outcomes of both monetary and nonmonetary nature.

The preceding decision problems involved changes of the status quo along a single dimension. Many decisions, however, concern transactions in which the possible outcomes include coordinated changes in several dimensions of value. The basic example of a transaction is the purchase of goods in exchange for money. Transactions are evaluated, we propose, according to the balance of costs and benefits in the relevant mental account. The framing of a transaction can alter its attractiveness by controlling the costs and benefits that are assigned to its account, as illustrated in the following problems.

> Imagine that you have decided to see a play where admission is $10 per ticket. As you enter the theater you discover you have lost a $10 bill. Would you still pay $10 for the play?

The great majority of subjects (88%) presented with this question said that they would buy a ticket for the play despite the loss of money. A second group of subjects was presented with the following problem:

> Imagine that you have decided to see a play and paid the admission price of $10 per ticket. As you enter the theater you discover that you have lost the ticket. The seat was not marked and the ticket cannot be recovered. Would you pay $10 for another ticket?

In this version of the problem, however, most subjects (54%) said that they would not buy another ticket. The marked difference between the responses to these problems is an effect of psychological accounting. We propose that the purchase of a new ticket in the second problem is entered in the account that was set up by the purchase of the original ticket. In terms of this account the expense required to see the show is $20, a cost that many of our respondents apparently found excessive. In the first problem, on the other hand, the loss of $10 is not linked specifically to the ticket purchase and its effect on the decision is accordingly slight. Indeed there is no reason in this case to post the loss of $10 to the theater-going account rather than to any other optional purchase.

Mental accounting is dominated by the tendency to group together the costs and benefits directly associated with an object or a project, as the following example illustrates:

> Imagine that you are about to purchase a jacket for $125 and a calculator for $15. The calculator salesman informs you that the calculator that you wish to buy is on sale for $10 at the other branch of the store, located 20 minutes' drive away. Would you make the trip to the other store?

Most respondents who answered this question gave a positive answer. Another group tackled a similar problem in which the cost of the jacket was changed to $15, and the cost of the calculator to $125 in the original store and $120 in the other branch. In this version, the majority of subjects gave a negative answer. The reader will easily confirm that the two problems have been equated in terms of the total purchase, as well as in terms of their immediate consequences. In both versions one has to decide whether to drive 20 minutes to save $5. The different responses to the two versions of the problem indicate that the respondents evaluated the saving of $5 in relation to the amount involved in the purchase of the calculator to which it pertained most directly. In relative terms, of course, a reduction from $15 to $10 is more impressive than a reduction from $125 to $120.

Framing effects in consumer behavior may be especially pronounced in situations that involve a single dimension of cost (usually money) and several distinct dimensions of benefit. The fancy tape deck is a distinctive asset in buying a new car, but its cost is naturally treated as a small increment over the high price of the car, and is evaluated as such. The purchase is made easier by evaluating its benefit on its own and its cost as an increment. Home-buyers frequently report similar experiences. Furniture is often bought with relatively little distress at the same time as the new house. Purchases that are postponed fortuitously, perhaps because the desired item was not available at the right time, often appear extravagant when contemplated separately, because their cost looms larger on its own. In all these examples, the attractiveness of a course of action may change if its cost (or its benefits) are embedded in a larger account.

If a decision is influenced by the reference point with which possible outcomes are compared, what determines the reference point? The dependence of impressions, judgments and responses on a point of reference is a ubiquitous psychological phenomenon. The same tub of tepid water may be felt as hot to one hand and cold to the other, if the two hands have been exposed to different temperatures. The same income may be considered lavish or inadequate, depending on whether one's earnings have recently increased or decreased. In these examples, the reference point is the state to which one has become adapted.

In many cases, however, the reference point is determined by events that are only imagined. Consider the following incident:

> Mr. Crane and Mr. Thomas were scheduled to leave the airport on different flights, at the same time. They traveled from town in the same limousine, were caught in a traffic jam, and arrived at the airport 30 minutes after the scheduled departure of their flights.
> Mr. Crane is told that his flight left on time.
> Mr. Thomas is told that his flight was delayed and just left five minutes ago.
> Who is more upset?

Almost everyone presented with this situation agrees that Mr. Thomas is more upset than Mr. Crane, although their objective conditions are identical: both have missed their flights. Furthermore, both should have expected to miss their flights. If Mr. Thomas is more upset, it is presumably because in the act of imagination they both perform, Mr. Thomas comes closer than Mr. Crane to catching his flight. The ease with which one can imagine a more desirable situation than the one that exists affects the frustration that is experienced in response to the unsatisfactory reality. For another example of the same notion, consider the following:

> The winning number in a lottery was 865304.
> Three individuals compare the tickets they hold to the winning number:
> Mr. A holds 361204;
> Mr. B holds 965304;
> Mr. C holds 865305.
> How upset are they?

There is a general agreement that the experience is worst for Mr. C, quite severe for Mr. B, and entirely mild for Mr. A. Here again, the ranking corresponds to the degree to which the individuals can be described as having "come close" to winning the prize.

An individual's experience of pleasure or frustration may thus depend on an act of imagination that determines the reference level to which reality is compared. Still, the act of imagination by which one creates alternative possibilities is not completely unconstrained. If there were no constraints, Mr. Crane would find it as easy to imagine himself catching his flight as Mr. Thomas does to imagine catching his, and Mr. A would find it as easy as Mr. C to imagine himself with the winning ticket. Imagination appears to be governed by rules, and the rules of imagination affect our experience of reality by controlling the alternatives to which it is compared.

Concluding Remarks

In summary, I have described in the course of this lecture several phenomena of judgment and choice which may be viewed as cognitive illusions. Although illusions and errors are but the medium by which some cognitive processes are revealed, the medium has influenced the message. The accumulation of demonstrations in which intelligent people violate elementary rules of reasoning and decision has suggested that some blemishes should be added to the portrait of human thinkers. They also raise some intriguing philosophical questions. How can a creature who is prone to so many errors and biases evolve and prosper? Why do we fail to learn the correct rules, or at least unlearn some of our erroneous beliefs? And finally how can we learn to avoid some of the errors and illusions to which we are prone and improve the quality of our decisions and judgments?

I will not attempt to answer these questions. I would only like to point out that the conditions under which people can infer rules from experience are not well understood. Some rules are learned almost without effort, while others seem impossible to learn. It would be premature, I believe, to conclude from the present findings that proper instruction in probability and decision theory will solve the problem. A perceptual example may illustrate the point. A dangerous illusion encountered while flying is that a pilot sometimes has a clear impression that the plane is ascending when it is in fact descending. In such circumstances, the pilot is not taught to revise the intuitive perception of motion. Rather he is told to trust the instruments.

The principles of logic, statistics and decision theory provide us with instruments against which we can check some of our intuitive decisions and judgments. Our ability to use such instruments wisely, I believe, can be improved by better understanding of our thought processes and the errors to which they are prone. The mark of intelligent behavior is not infallibility but rather the ability to recognize mistakes and learn from errors.

References

Gregory, R.L., *The Intelligent Eye* (New York, St. Louis, San Francisco: McGraw Hill, 1970).

Kahneman, D., Slovic, P., and Tversky, A. (eds.), *Judgment under Uncertainty: Heuristics and Biases* (Cambridge Mass.: Cambridge University Press, 1982).

Kahneman, D. and Tversky, A., "On the Study of Statistical Intuitions," *Cognition*, III (1982): 123–141.

Shafer, G., *A Mathematical Theory of Evidence* (Princeton: Princeton University Press, 1976).

Tversky, A., "Features of Similarity," *Psychological Review* 84 (1977): 327–352.
Tversky, A. and Kahneman, D., "The Framing of Decisions and the Psychology of Choice," *Science* 211 (1981): 453–458.
Tversky, A, and Kahneman, D., "Extensional versus Intuitive Reasoning: The Conjunction Fallacy in Probability Judgment," *Psychological Review* 90 (1983): 293–315.
Zadeh, L.A., "Fuzzy Sets as a Basis for a Theory of Possibility," in: *Fuzzy Sets and Systems,* 1978, pp. 3–28.

The Past of an Illusion
A Comment

AVISHAI MARGALIT

The juxtaposition 'cognitive illusion' is interesting and important. Firstly, I shall try to say why it strikes me as interesting and what's in it that is of importance, by framing it within a historical context. Secondly, I shall address my comments to one alleged cognitive illusion, namely the conjunction fallacy.

I

The expression 'cognitive illusion' is not to be contrasted with 'perceptual illusion'. It contrasts instead with 'motivational illusion'. Indeed, the whole point of adducing 'illusion' to 'cognition' is to place it on a continuum with perceptual illusions of the stick-bent-in-water kind. In both cases the seductive force of the illusion remains even after we recognize that it is an illusion. Knowing that the stick is in fact straight, we still see it as bent. In traditional discussions on errors and mistakes, the central idea is that our illusions and mistakes are due to motivational sources. Our cognitive equipment, according to this view, is good as it is as long as it is not tampered with by external forces. (External to cognition, that is, yet internal to our psychological makeup.) Factors external to cognition are things such as drives, wants, wishes, emotions, and feelings; any of these may derail cognition from its rational course.

89

E. Ullmann-Margalit (ed.), The Kaleidoscope of Science, 89–94.
© *1986 by D. Reidel Publishing Company.*

The novelty in the idea of 'cognitive illusion' is in locating the source of illusions in the very cognitive mechanism itself. The progenitor of this idea is Kant. He even reserved a special use of the term 'dialectics' for designating the science of illusions. We shall come back to Kant later.

The idea that error is a proper topic for philosophical reflection was conceived at the time philosophy itself was born. One question that we inherited from Plato, who in turn had inherited it from Parmenides, is: "What is an erroneous sentence about?" Since error reflects a mistaken belief, then, if a true belief is about some fact in the world, what is a false belief anchored in? In nothing, says Parmenides. If so, then to err is to believe in nothingness. This of course is senseless, hence the notion of error itself must be senseless. But we obviously do err. Hence a puzzle.

Another old riddle about errors is more pertinent to our concern with cognitive illusions. Descartes was the one who put it on the philosophical agenda, though with a theological twist. The problem he posed is: Why should the good God have instilled in us the possibility of error? Error is bad; and so an omnipotent and benevolent God should have done better than equip us with the unrealiable cognitive mechanism we seem to have. Our errors thus challenge either God's omnipotence or His benevolence. Perhaps even His omniscience.

We can, however, remove God from the scene and pose the problem in evolutionary terms. The question then is: How can an ingrained illusion in our system of cognition be explained in terms of natural selection? We would surely be far more adaptable to the world without our crippling illusions; so why did nature, throughout the years, not filter out our entrenched illusions, perceptual as well as cognitive? Or perhaps such illusions do have their adaptive role after all?

Descartes' answer is well known. Our cognitive apparatus is good as it is. Following the right method yields, in his view, the right results. Our will is what gets in the way of correct cognition. Mistakes are the bastards of Cognition raped by the Will. The Will is free, this is what makes us free agents, but not free from making mistakes. Descartes' Will is what we nowadays call "motivation": the idea is one and the same. Mistakes, errors and illusions are due to the interaction between our cognitive mechanism and our other psychological mechanisms. Our frictionless cognitive competence is free of errors, but our resultant performance is full of them.

This, I contend, was, and perhaps still is, the dominant picture in psychology and philosophy. Tversky and Kahneman are not the first to challenge this picture, but they have perhaps done more to free us from its grip than anyone else.

There are, to be sure, motivational illusions. We find them even in the realm of perception. This we know e.g. from the experiments of Bruner and Goodman[1] in which children from poor families tend to overestimate the size of coins more than children from well-off families. From the experiments of McClelland and Atkinson[2] we learn that food objects were estimated as larger by subjects who were hungrier. If perception is affected by motivation, all the more are our other faculties liable to be so affected. In love, says Nietzsche, our power of illusion reaches its zenith. And Freud of course hammered the point that the sources of many of our illusions are instinctive drives. According to Freud, illusions are often due to a futile effort to control the external world by wishful thinking. Thus the idea of God as a substitute for the slaughtered patriarch is for Freud a clear example of how illusions are motivational. His view is markedly different from that according to which God is seen as a product, or an outcome, of our internal cognitive tendencies to create ever larger unifications, even such that transcend experience. According to this last view, God is also an illusion — a cognitive illusion.

For Kant, who is perhaps the originator of this view, metaphysics in general is a gigantic cognitive illusion. This view is shared by some of the most influential philosophical trends in our century. The metaphysical grand illusion is, according to Kant, an inescapable tendency — like the unavoidability of seeing the sea, viewed from the shore, as higher in its center. Kant drew the deep analogy between perceptual illusions and cognitive illusions. He had some telling explanations as to what in our cognitive system is responsible for the metaphysical illusion. This happens, for example, when heuristic principles (in his language, 'regulative rules') are taken as constitutive principles of reality, that is, when we project on reality features which belong not to that which is represented but to our way of representation. This idea of Kant's became a battle-cry of various schools of analytical philosophy. All this is, admittedly, very sketchy indeed, but it is still a picture worth drawing; as from Kant on there is a trend to locate the source of many illusions in our cognitive structure.

This is the context in which Tversky's paper should, I believe, be taken.

In the view of Tversky and Kahneman, our mistakes in judgment are not deviations from our Bayesian true selves due to wishful thinking. They are, rather, the outcome of the very heuristics that underlie our probability judgments. True, our mistakes in judgment are considered to be mistakes when judged from the perspective of a normative theory. But the normative theory is not a description of our true nature clouded by irrational feelings. The old-fashioned division of mental labor between reason and feelings is not to Tversky's liking.

II

Tversky, Kahneman, Maya Bar-Hillel and others have suggested that there are *systematic* deviations of people's probability judgments from normative prescription. They also demonstrated that this is the case not only with naive subjects but with sophisticated ones as well. There are thinkers who disagree. Perhaps the most persistent of these is Jonathan Cohen. He claims that though the judgments assessed by Tversky and Kahneman do deviate from one normative calculus (which he calls the 'Pascalian' calculus), they nevertheless cohere with another normative calculus which he himself developed (a 'Baconian' one).

The force of the conjunction fallacy discussed by Tversky is that no one in his right mind will suggest a calculus in which the probability of a conjunction is greater than the probability of one of its conjuncts. Thus if Tversky is correct about the conjunction fallacy, he has a knock-down example in his possession. However, although I join Tversky and Kahneman in the belief that there is in fact a conjunction *effect* of the kind they suggest, it does not seem to me that easy to discern it amidst its background noise.

Let me try to spell out an alternative explanation for the fallacy involved and thereby demonstrate how difficult it is to end the rounds of the controversy with a knockout.

Piaget, you may recall, pointed out that children under the age of seven or eight find it difficult to answer correctly questions which relate to class inclusion. Thus Piaget shows to the six-year-old Staro a drawing of 15 buttercups and two bluebells. "What are those?" — "Buttercups." "And those?" — "Bluebells." "Are they all flowers?" — "Yes." "Are there more flowers, or more buttercups?" — "More buttercups." "Why?" — "There are only two bluebells." "But aren't the buttercups flowers?" — "Yes." "Then are there more buttercups or more flowers?" — "More buttercups."

With respect to probability judgments of conjunction we are perhaps in a situation not unlike Staro's. An explanation of Staro's way of computation, in the spirit of Tversky and Kahneman, goes like this: Staro works with the intensions of 'buttercups' and 'bluebells' rather than with their extensions. True, the set of flowers (extensions) includes the set of buttercups. But the set of buttercup properties (intensions) includes the property of being a flower. Children under seven, so the explanation goes, find it hard to think of sets extensionally; they view them through their representative properties instead. And since there is an inverse relation between extension and intension inclusions, the child mistakes one for the other.

Quine once remarked that philosophers are psychologists without research grants. Accordingly, I have not myself asked Staro and his likes if there are more bluebells or more flowers in the drawing. My hunch, however, is that if I had asked them whether there were more *bluebells* (remember: there were only two of those) or more flowers, the answer I would have received would be "more flowers." If I am right in my hunch then the explanation in terms of intension inclusions breaks, since the intension of 'bluebell' also includes that of being a flower. An alternative explanation is this: the children do not make a mistake in computation; rather, they commit a lingual error. When asked whether there are more buttercups than flowers, they understand the question with a twist, namely, whether there are more buttercups than *other* flowers. 'Buttercups' and 'flowers' are taken by Staro as designating complementary sets, not containing and contained ones.

I suspect that something like this happens in the conjunction case. Let us look again at the example used by Tversky.

    (1)   Bill plays jazz as a hobby.
    (2)   Bill is an accountant and plays jazz as a hobby.

Now I suggest that (2) is not grasped, in probability judgments, as a conjunction, but is taken as an implication: What is the probability that Bill's hobby is playing jazz *given that* he is an accountant, as compared with the simple probability that this is his hobby. Against the background of the provided description of Bill's personality, the answer is that it is more likely that he is a jazz-playing accountant than that he is a jazz-playing something else. In short, what is judged as the unlikely event in the conjunction sentence (namely, Bill's playing jazz) is taken as an antecedent, not as a conjunct.

Much the same holds for the case of Connors and McEnroe. It is possible that the subjects understand the sentence "McEnroe will lose the first set and win the game" as: "(Even) if McEnroe loses the first set he will (still) win the game." And since the subjects judge McEnroe to be the better player of the two, they assess that probability as higher than the probability of McEnroe's simply losing the first set.[3]

Finally, a question. If, as Bishop Butler urged us to think, induction is a guide to life, how come God, or for that matter Evolution, did not provide us with a better guide? I do not have an answer, but would like to suggest an analogy which may help. (It is Carnap's parable put to a different use.) Consider a Swiss pocketknife with numerous blades. This knife can be used for a large variety of purposes and activities, even though for each given activity — say, slicing bread, clipping fingernails, etc. — there is a special instrument which is more efficient. So the pocketknife, while relatively crude for each activity, is yet the best instrument for performing *all* of them. Now the scalpel of normative probability calculus is certainly better than the blade óf probability judgments by representativeness. But judgments of representativeness serve us in a variety of tasks: classifications, "computations" of meaning, and also in probability judgments. For these many services we pay a price. Illusions are the price. This is the price we pay for being pocketknives with many blades rather than scalpels. It may, after all, be a price well worth paying.

Notes

1 J.S. Bruner and C.C. Goodman, "Value and Need as Organizing Factors in Perception," *Journal of Abnormal Social Psychology* 42 (1947): 33–44.
2 D.C. McClelland and J.W. Atkinson, "The Projective Expression of Needs: The Effect of Different Intensities of the Hunger Drive on Perception," *Journal of Psychology* 25 (1948): 205–22.
3 There is yet another way to explain away the conjunction fallacy, by claiming that subjects in our case reverse the procedure. Instead of estimating the probability of Bill's being an accountant (or a jazz-playing accountant) on the basis of his given character description, they judge the likelihood of his character description on the basis of his being an accountant (or a jazz-playing accountant). According to this interpretation, then, people often judge what is technically called "likelihood" rather than probability. And if so, then the so-called conjunction fallacy is no fallacy. I deal with this interpretation of the conjunction fallacy in my "More than Likely," *Times Literary Supplement*, August 26, 1983, p. 914.

Molecular Genetics and the Falsifiability
of Evolution

BERNARD D. DAVIS

In 1982 in the United States the American Civil Liberties Union won a legal battle against a state law requiring teaching of the creationist view of man's origin along with any teaching of evolution in public schools. But despite this victory it is clear that skepticism about evolution remains a threat — and not only in the United States — to efforts to build our understanding of man on scientific insights into reality.

The major reason for the resistance, of course, is the conviction that society cannot build a satisfactory moral consensus without a transcendental foundation, and the fear that the evolutionary view of man's origin destroys that foundation while failing to provide an alternative. In addition, in recent years the left, despite its materialist orientation, has also exhibited resistance to evolution: not to its implications for man's origin, but to its implications for human genetic diversity. With this group the basis is the conviction that the pursuit of equal opportunity depends on faith in the virtual equality of potential, or at least is chilled by lack of that faith.

These charges reflect sincere concern over serious problems, to which science has clearly contributed. But when myths and faith come into serious conflict with reality, it seems to be a matter of definition that reality will have the last word. Moreover, while scientific insights into our origins and nature are a relatively recent addition to man's history, they are bound to continue

E. Ullmann-Margalit (ed.), The Kaleidoscope of Science, 95–110.
© *1986 by D. Reidel Publishing Company.*

to grow, and it is hard to see how they can fail to prevail in the long run. I would therefore like to consider how recent developments in biology can help us to present the evidence for evolution in a way that a wide public might find more convincing than the traditional emphasis on the fossil record.

Let me start with a philosophical argument of Karl Popper, which highlights the limitations of the traditional approach.

Popper, Darwin, and Modern Evolutionary Theory

Drawing on the laws of physics, Popper made a major contribution to the philosophy of science by proposing a sharp line of demarcation between scientific and metaphysical theories: only the former can generate testable, falsifiable[1] predictions.[2-4] But when applied to biology this criterion created a dilemma: for while Popper accepted evolution as a fact, the strict application of his criterion led him to conclude that Darwin's theory, even in its modern version, is a metaphysical research program and not a scientific theory. To be sure, for Popper 'metaphysical' was not a pejorative term. Moreover, he later conceded that his criterion was too rigid for those sciences that include a historical element.[5,6] Nevertheless, creationists have been able to draw support from his original position.

In fact, since Popper has written so brilliantly on other matters, it is surprising that he is willing to make broad pronouncements about the nature of biology while displaying remarkably limited knowledge of the field and little understanding of how its advances really arise. I largely agree with the severe critique of Popper's views on biology presented by a philosopher, Michael Ruse.[7] To summarize his main points: Popper projects his own intellectual preoccupations onto Darwinism, reading features of the evolution of knowledge into the evolution of organisms; his view of science is prescriptive and not descriptive; he does not appreciate the radical change that genetics made in evolutionary biology, or the sophistication of the field that has resulted; in his effort to "enrich" this "feeble" theory by a sort of behavioral saltationism, in which behavioral adaptations precede anatomic changes, Popper reflects a philosopher's obsession with mind and ignores the world of microbes and plants, which are equally important for evolution; he appears to be unaware of the importance of populational diversity in biology, in contrast to the homogeneity of each class of entities in physics; and in dismissing natural selection as a tautology he fails to recognize the empirical depth of modern evolutionary theory (see below). As a biologist I would further add that Popper, concerned with making the logic of science rigorous, fails to recognize how much the progress of biology has depended on two features that do not lend themselves to logical prediction: the serendipitous observations that open up previously unrecognized or inaccessible problems; and the development of novel techniques for analyzing or manipulating increasingly complex materials. Both points, for example, are vividly illustrated by the dramatic development of the recombinant DNA technology.

Accordingly, while Popper claimed to be analyzing the present state of evolutionary biology and its failure to be built on testable predictions, he clearly has assimilated very little of the maturation of that field since Darwin. Moreover, even the theory as presented by Darwin contained large elements that can be formulated as testable predictions — for example, the correlation between the age of fossils and their complexity, or the presence of parallel sets of species, filling similar ecological niches, in geographically separated regions.

Nevertheless, if Popper were considering only the theory as presented by Darwin he would have real grounds for his dilemma: for that theory does differ radically from modern, experimental science. Since Darwin created the theory before the emergence of genetics, he had no direct evidence, or even a plausible mechanism, for one of its major components: the postulate of continuing hereditary innovation. Accordingly, instead of being built, in stepwise fashion, on a continuous sequence of firm discoveries and predictions, the theory required a large conceptual leap.

Darwin's theory was thus in a sense premature. For this reason it remained controversial among biologists until the 1930s, when the so-called 'modern synthesis' of evolution with genetics filled the gap and demonstrated evolutionary shifts in populations. The model of evolution emerging from this synthesis, called neo-Darwinian, then became the central, unifying principle of biology. Nevertheless, a skeptic could still correctly claim that the idea of macroevolution, connecting all organisms in one branching taxonomic tree, was still an inference from only indirect, though highly coherent, evidence, based on comparisons of present morphological phenotypes and on records of the historic past. But today the picture has changed radically. Molecular genetics now provides extraordinarily direct evidence for the continuity of the whole range of existing organisms. Yet in the recent trial over the creationist law this evidence received essentially no attention.

Of course, the sophisticated molecular evidence can hardly be expected to persuade a diehard fundamentalist. But in the teaching of evolution to the next generation this evidence surely deserves prominence, alongside the classical evidence and the recognition of Darwin's genius. For the new, direct evidence is likely to be most convincing to a student, just as the view from a satellite is now the easiest way to convince a child that the earth is round. I shall therefore review how the successive stages in the growth of genetics, up to molecular genetics, have progressively strengthened evolutionary biology.

The Problem of Variation

Before starting our review, let us consider in a little more detail Darwin's problem with variation. His theory had two equally important components: hereditary variation and natural selection. The latter received all the attention, not only because of its great scientific novelty and significance, but because its courageous rejection of a finalistic cause had deep philosophical and religious implications. Nevertheless, the idea itself, summed up in the phrase "survival of the fittest," is logically so inevitable that by itself Popper would be justified in considering it a tautology. What makes the notion of selection really significant is its linkage to two postulates: that the existing differences in fitness are inherited, and that they can undergo inheritable changes. This is the scientifically radical part of Darwin's theory. But he could say very little about the processes involved.

To be sure, in invoking hereditary variation underlying natural selection Darwin did lean heavily on the model of artificial selection in domesticated animals and plants, as practiced by his neighbors in the English countryside. But that model has a serious weakness. We now know that the variation used in artificial selection arises largely by reassortment of the genes already existing within a species; and without a science of genetics, one could imagine that it might have arisen entirely by that mechanism. In contrast, if the enormously diverse living world had arisen from a common ancestor, a much deeper kind of genetic innovation would be required. Darwin thus had to postulate a pattern of heredity with apparently contradictory properties: breeding true, and yet continually generating real novelty, i.e. variation beyond the range already present in the parents or even in the species. Wrestling all his life with the problem of defining a plausible mechanism, Darwin considered 'hard' inherited variation, independent of the environment, the main source; but commonsensical observations prevented him from discarding the Lamarckian notion of 'soft' inheritance, responsive to use and disuse.[8] This mixed view led him to a logical but useless theory of 'gemmules' — particles of inheritance released from any body cells, responsive to use and disuse of those cells, and able to reach the germ cells.

Even if Darwin had known of Mendel's discovery, contemporaneous with his own, he would have found no help; a much more sophisticated science of genetics was required. To appreciate the contribution of this growing science, let us imagine how genetics and evolution might have developed in a logical sequence — that is, let us pursue a hypothetical scenario, based on the fanciful assumption that no one dared to propose the theory of Darwin and Wallace, or at least that no one was able to sell it, until it had a testable foundation in genetics.

A Hypothetical Scenario

The first step toward filling Darwin's major gap was DeVries' discovery of mutations, in 1900. Hereditary variation could then be seen to arise in two different ways: mutation provided the ultimate source of novelty, and reassortment of genes by genetic recombination enormously amplified that novelty. Nevertheless, the impact of genetics on evolution was long delayed. For decades geneticists, on the one side, believed that species arose by giant mutations ('saltations'), rather than by the gradual changes invoked by Darwin; while evolutionists, on the other side, disregarded mutations as exceptional monstrosities and assumed that the gradual steps of evolution arose by some other mechanism. The problem was solved, in the 1920s, by the recognition that traits exhibiting continuous hereditary variation have a fundamentally 'particulate' basis, just like the discontinuous traits of Mendel: the apparent continuity arises from the additive small contributions of many genes to the same trait (polygenic inheritance), as well as from the interactions of these genes with the environment and with each other. The mathematical contributions of Fisher, Haldane, and Wright provided a solid basis for this view, and the development of population genetics then provided a major new tool to evolutionary biology.[9]

By this time, in our hypothetical scenario, it would be clear that the hereditary process can generate an endless supply of genetic novelty, in all classes of traits. Study of the resulting diversity led to a profound change in our conception of the nature of biological populations. Earlier biologists, under the influence of Aristotelian essentialism, had long viewed species typologically (i.e. in terms of a normal or ideal type, with only minor variations among individuals). Accordingly, early geneticists thought primarily in terms of wild-type (normal) and mutant alleles of a gene. However, population geneticists demonstrated widespread polymorphism (naturally occurring multiple forms of a gene within a species). It was then realized that each species contains a great deal of genetic diversity, which the sexual mode of reproduction conserves and reshuffles. As Mayr[10] has emphasized, the shift to this populational view, recognizing the statistical diversity of natural populations, has been one of the most important conceptual advances in biology.

In addition to demonstrating this diversity, the development of population genetics also made it possible to demonstrate microevolution directly. For example, in localities where industrialization darkened the tree trunks, an originally light-colored species of moth was found to be replaced by dark variants, within relatively few generations; and while such a shift would have been ascribed earlier to physiological adaptation, the difference was now found to be hereditary (genetic adaptation). In addition, population geneticists recognized the true nature of geographic races: within both animal and plant species, races are populations that have accumulated significant statistical differences in their gene pools, as a result of prolonged reproductive separation. With this insight it was also recognized that races could serve as incipient species: reproductively separated groups whose progressive divergence (e.g. in behavior, genital fit, and chromosomal organization) would eventually give rise to separate, reproductively incompatible species.[10,11] Finally, the development of theoretical population genetics provided an elaborate mathematical basis for calculating the contributions of the many factors that affect the kinetics of evolution, including selection pressure, ecological variation, mutation rate, migration rate, assortative mating, and random drift arising from the isolation of small populations.

Incidentally, since the alleged lack of predictive power is what disturbs Popper most about evolution, we might note that the mathematical character of modern population genetics implies strong predictive power. In addition, the discovery of genetic drift, as exceptions to the survival of the fittest, further removes the tautology from evolution.

The microevolution that had been a bold hypothesis thus became a description of fact, just as Harvey's bold hypothesis of blood circulation became a description when capillaries were discovered. The idea of macroevolution, creating the enormous diversity of the living world from a single origin, did not have as firm a base; it still rested largely on graded morphological homologies, past and present. However, with the evolutionary interpretation of these homologies so strongly supported by genetic principles and by the observation of microevolution, evolution became the central organizing principle of biology.[8,12,13] Meanwhile, the morphological homologies were increasingly supplemented by another kind of evidence for macroevolutionary continuity, from the growing science of biochemistry.

The Unity of Biochemistry

Macroevolution leads to a prediction of conserved unity along with emerging diversity: even after extensive divergence, organisms might be expected to retain some common features. And while early studies in biology, on gross structure and function, focused on its still astounding range of diversity (who can fail to marvel at the complex elegance of the peacock's feathers or the chambered nautilus?), as biology penetrated into smaller dimensions it provided increasing evidence of conserved unity. Even in Darwin's time the microscope had revealed common structural features of all cells. Later, the development of comparative biochemistry, from the 1930s on, provided detailed evidence for much additional conservation, at a molecular level: cells of the most distant organisms, ranging from bacteria to man, use the same building blocks and metabolic pathways, with remarkably few exceptions. More recently, molecular genetics has demonstrated unity at an even deeper level: in translating genetic information into protein sequences all organisms use the same genetic code (except for minor variations in sequestered organelles) and they use essentially the same machinery.

Meanwhile, the study of biochemical regulatory mechanisms revealed an unexpected additional kind of unity. The binding of a small molecule to a macromolecule (e.g. a substrate to an enzyme) ordinarily shows hyperbolic kinetics: with increasing concentration of substrate, the binding increases less and less, as it approaches saturation. Hemoglobin, however, has a sigmoid oxygen dissociation curve, i.e. with increasing oxygen concentration its binding rises slowly, then rapidly, then slowly again. The advantage is clear: the uptake of oxygen in the lungs and its delivery to the tissues requires that the latter have a lower oxygen tension, and the sigmoid curve of hemoglobin, compared with a hyperbolic curve, allows a much larger delivery for the same difference in tension. Physiologists had long regarded this property of hemoglobin as a marvelous special product in the evolution of vertebrates. Much later, however, the primitive, single-celled bacteria were found to have evolved fundamentally the same phenomenon, for parallel purposes: the activity of the initial enzyme of each biosynthetic pathway is regulated by the concentration of the end-product of the pathway, which combines reversibly with the enzyme protein and shifts it from an active to an inactive shape (called an allosteric response); and the kinetics of the response is also sigmoid, thus adjusting the rate of synthesis to maintain a constant concentration of the end-product in the cell. This response utilizes the same molecular mechanism observed for hemoglobin: the protein contains multiple subunits, and these exhibit cooperative conformational interactions. That is, binding of a molecule of oxygen (or of an end-product) by one subunit in the complex changes the shape of an adjacent subunit, in a way that increases its affinity for the next molecule.

Though it was a surprise to discover that single-celled organisms had evolved such an elaborate mechanism, in retrospect we can see that pressure for faster and more efficient growth must have selected early in evolution for such feedback of regulatory information, economically linking supply and demand. We will consider later the broader evolutionary significance of the phenomenon of allostery.

The Impact of Bacterial Genetics

Curiously, the genetics of bacteria developed very late, because these small cells long seemed to lack the elements required for genetic studies: visible chromosomes, mutable genes, and mechanisms for recombining genes. Bacteria were therefore regarded as virtually bags of enzymes, with some vague, plastic kind of inheritance. However, in the 1940s discrete mutations were discovered in bacteria, and mechanisms of gene transfer soon followed. These organisms were then rapidly absorbed into the elegant framework of genetics, and hence of evolutionary biology. Moreover, the very features that had impeded the analysis of their variation then became advantages: simpler cellular structure, smaller genomes, rapid multiplication, and efficient selection of rare mutants and rare recombinants from huge populations. Bacteria and their viruses then became the model organisms in the revolution that identified the formal units of genetics as molecular sequences. One key step was Beadle and Tatum's discovery, in 1940, of the 1:1 relation between genes and enzymes in the mold *Neurospora* — a discovery soon extended to bacteria. Avery's analysis of pneumococcal transformation, in 1944, was even more important. By identifying the material of the gene as DNA, and also disclosing gene transfer in bacteria, it opened up two fields: molecular genetics and bacterial genetics.

And now a word about bacteria and evolution. The discovery of gene transfer and recombination in the simple, haploid prokaryotes (bacteria) came as a surprise. Yet in retrospect we can easily understand why genetic recombination appeared in evolution long before it became a regular mechanism, in the sexual reproduction of the more complex diploid eukaryotes. For the large step from prokaryotes to eukaryotes, after half of the 4 billion years of biological evolution, clearly required a large accumulation of genetic variation. If the only source of this novelty were successive mutations within a lineage of prokaryotes, without any form of recombination, it is hard to imagine that evolution would even yet have progressed beyond the prokaryotes.

The development of bacterial genetics not only provided simple model cells for pursuing molecular genetics. It also added the enormous diversity of the bacterial kingdom to the range of organisms studied in evolutionary biology. In addition, bacteria led to a new dimension in evolutionary biology: serious study of the origin of life. That problem used to be seen in terms of a specific, dramatic event, in which the organic molecules accumulating in a prebiotic "soup" eventually formed a self-replicating aggregate — the first cell. What we now know about genes makes it much more reasonable to postulate a process of gradual prebiotic evolution, in which genetic information was encoded in nucleic acid sequences that grew in precision of replication, and hence in the complexity of the information that they could transmit. The increasingly rich detail in this model, and the increasing experimental evidence for many of its postulated features, ought to win round those who would deny Darwinian evolution because it lacks a plausible undirected beginning.

Molecular Mechanisms of Genetic Variation

One major contribution of molecular genetics to evolution has been a great increase in our understanding of the nature of mutations. Watson and Crick's second paper, in 1953, noted that the double-structure of DNA could explain them as occasional errors in base-pairing. And while that recognition of a molecular basis for mutations itself did not immediately advance evolutionary biology, it did lead to a later interesting prediction: given the mechanism of DNA replication, in which one strand serves as template for the synthesis of the complementary strand, a process free of errors is thermodynamically impossible. Accordingly, occasional mutations not only are possible but are inevitable, and so mutability is not an evolved feature of gene replication but is an inherent property. In the early stages of prebiotic evolution, the problem was not to provide for alterations but to reduce them to a tolerable level, so that a useful sequence could be perpetuated.

Once the molecular study of mutations began, further work revealed an extraordinary (and still growing) variety of additional mechanisms. Because these range widely in the magnitude of their effects, they provide great flexibility to the steps underlying the Darwinian process of gradual change. In particular, since the extrapolation from microevolution to macro-evolution has often been challenged, it is of interest to note several mechanisms by which a single change in DNA (a 'mutation' in the broad sense) can produce a large effect on the products of translation. Let us consider four.

First, a single-base change in a sequence can alter several products — for example, if it occurs in a regulatory protein in a DNA sequence that regulates several adjacent genes, or in an enzyme that can modify a set of other proteins.

In a second set of mechanisms, in higher organisms, the processes of cell differentiation and embryonic development offer large opportunities for amplifying the effects of a small genetic change. We are barely beginning to understand the mechanisms, in which complex regulation of DNA appears to be the major factor.

A third class of mutations with a broad effect are chromosomal rearrangements. Rare microscopically visible rearrangements were discovered in higher organisms decades ago, but more recent molecular studies in bacteria have shown that smaller rearrangements are even more frequent. Their functional effects include an increase or decrease in the expression of a gene, or its inactivation, or creation of a novel protein by fusing parts of different genes. Moreover, bacteria frequently contain specific transposable (mobile) DNA sequences, readily transferred from one site to another through the action of special enzymes that recognize special sequences at their termini. The evolution of such a special mechanism shows that DNA rearrangement must be an important feature of evolution.

A fourth mechanism that produces large effects in one step is infectious heredity: the introduction of a block of exogenous DNA into the cell, and then often into one of its chromosomes. This process is prominent in bacteria but has also now been observed in animal cells. It is mediated most efficiently by viruses, and in bacteria also by the small accessory chromosomes called plasmids (which often are transferred by conjugation through a bridge between cells). Both of these agents of transfer not only can introduce their own genome but can also bring in an attached block of DNA picked up from an earlier host. In addition, cells can take up naked DNA, whether a chromosomal fragment (transformation) or a viral genome or a plasmid (transfection).

The ability of viruses to exchange DNA with the host cell not only suggests that they have evolved by breaking away from chromosomes; it also suggests a new conception of their role in nature. Just as the role of bacteria in disease is clearly a sidetrack from their much larger geochemical role in recycling organic matter to the atmosphere, the role of viruses in disease may well be only a sidetrack from a more important evolutionary role in mediating gene transfers.

Additional Predictions from Molecular Genetics

Returning to our concern with falsifiability, we might note that discoveries in molecular genetics have also met several more specific, testable evolutionary predictions. First, the emergence of increasingly complex organisms would require mechanisms for expanding the genome, as well as for mutating genes. We have already noted two such mechanisms. One is the insertion of additional DNA, acquired by infection from the outside. In the second mechanism, two chromosomes cross over (recombine) not as usual at homologous sites, yielding two complete chromosomes; they cross over at nonhomologous sites, so that the sequence between the sites is deleted from one product and duplicated in the other. When a gene is thus duplicated within a chromosome, one member of the pair can remain unchanged, to maintain the original (usually essential) function, while successive mutations in the other can create a new function.

Another prediction is that the mutation rate per unit length of DNA should vary inversely, at least roughly, with genome size: a rate that is appropriate for evolution in the small genome of a virus would create excessive frequency of mutations, mostly deleterious, in the 10^4-fold larger genome of a vertebrate. In fact, this prediction has been confirmed, and two molecular mechanisms have been discovered: variations in the accuracy of the enzymes that replicate DNA, and variations in the activity of the enzymes that correct errors in that replication. In addition, there are also mechanisms that affect the mutation rate at specific sites in the DNA within an organism. Evolution can thus select for higher or lower mutability, general or localized.

Finally, because genes are sequences in a DNA continuum, rather than discrete molecules, the multiple reading of a sequence is physically possible, and one could predict that the resulting economy would have evolutionary value, especially in the small genome of a virus. In fact, three mechanisms are predictable, and all have been found in viruses in recent years: the two strands are read independently, in opposite directions; a strand is read between different starting and stopping sites; and, because of the triplet genetic code, a given DNA sequence can be read in three different phases, yielding three different products.

Direct Evidence for Macroevolution: Sequence Homology

As we have seen so far, microbial genetics and molecular genetics have given us deep insights into the details of the mutational process, revealing a much wider variety of mechanisms than had been envisaged earlier. But in an even greater contribution molecular genetic studies on a variety of presently living organisms have demonstrated their evolutionary linkage quite directly, through novel evidence of two kinds.[14]

The first, and most extensive, body of direct evidence has been based on a strong prediction: if species diverge in structure and function through the accumulation of changes in DNA (and hence in protein) sequences, then differences in these sequences should parallel the phenotypic differences and should be a direct measure of evolutionary distance. This approach, creating the new field of molecular evolution, has blossomed as techniques have been developed for comparing sequences: first the sequences of proteins; then measuring the overall similarity of the DNA of different species by melting apart their strands in solution and then measuring their formation of hybrid double strands at annealing temperatures; and now the development of simple techniques for determining DNA sequences. It is now possible to count the precise number of differences between homologous genes in two different organisms, and also to measure the total frequency of differences in sequence (polymorphism), with or without phenotypic differences, within a species.

The results have generally confirmed the branching taxonomic trees resulting from earlier approaches in systematic biology, while the few discrepancies provide a challenge. In particular, the close evolutionary connection of the human species to primates is now supported by brutally direct evidence: the average sequence homology between human and chimpanzee is about 99%,[15] compared, for example, with 70% between mouse and rat.

In addition to providing measurements of evolutionary distance, these approaches have also opened up a host of new insights and challenges. For example, higher organisms possess much DNA whose function is not known, including enormous amounts of repetitive DNA (sequences repeated many thousands of times) and 'pseudogenes' that are almost identical with active genes but are inactive. Moreover, gene expression in higher organisms requires transcription into RNA sequences in the nucleus, which then are processed to yield the messenger RNA that directs protein synthesis in the cytoplasm; but some genes now appear to have been formed by reverse transcription from messenger RNA.

We do not yet know why these several unexpected classes of sequences have accumulated: whether as inefficiently eliminated residues of once active genes, as elements selected for yet unrecognized functions, or as 'selfish DNA' that is amplified because the mechanisms responsible for DNA replication simply cannot prevent such opportunism. But we do know that a whole new perspective has been opened up in evolutionary biology. Evolutionary changes in DNA will have to be accounted for not only in terms of their effects on the phenotype of the organism but also in terms of their effects on the properties of the DNA itself — for example, its stability, its interactions with plasmids, viruses, and restriction enzymes, its tendency to undergo rearrangements and extra replications, its tendency either to be more firmly base-paired in a given region or to 'breathe' more readily (because of differences in content of the A-T and the G-C base pairs), and its content of sequences that can give rise to intrastrand loops by base-pairing within each strand.

Further Evidence for Macroevolution: Interspecific DNA Transfer

Let us now consider a second kind of compelling evidence that molecular genetics provides for macroevolution: the transfer of DNA between cells of distant species, not only in the laboratory but also in nature. For example, tumor viruses can pass between different animal species and then be integrated into their chromosomes; crown galls develop in plants by integrating genes from a bacterial plasmid into the host cells;[16] and a luminescent bacterium symbiotically associated with a fish has evidently taken up a gene from its host, since it has the characteristic animal form of the enzyme superoxide dismutase (containing Cu and Zn) rather than the quite different form (containing Fe) found in other bacteria.[17]

Compared with these transfers of DNA between distant organisms, transfer between closely related organisms is even more likely to yield an effective product and hence to survive in evolution. However, the products would then be harder to recognize as distinctive, compared with transfers between distant species. Fortunately, DNA sequencing offers an alternative method for detecting such lateral gene flow. Thus, in a pair of sea urchin species long separated in evolution, the sequence divergence in one histone is less than 1/100 of that observed in other proteins, including other histones; hence these organisms seem to have shared this gene quite recently.[18]

It seems clear that of all the DNA transfers occurring between species, only a small fraction will become established in evolution, and most of these are likely to be difficult to detect. Nevertheless, with the evidence at hand we must infer a continual slow lateral flow, reinforcing, as it were, evolutionary connections that were established much earlier by the process of branching vertical flow.

Incidentally, this finding of interspecific DNA transfer in nature also has interesting implications for the recent controversy over the hazards of recombinant DNA. Since fragments of foreign DNA have clearly been finding their way into bacteria in nature for a long time, the recombinant bacteria now being produced in the laboratory are not so novel, as a class, as was initially believed. The human species must therefore have survived continual exposure to random recombinants (for example, intestinal bacteria that had taken up DNA released from adjacent host cells). If this evolutionary principle had been more widely recognized, we might have been spared much of the anxiety that arose over the presumed danger from such organisms.

Conclusions

In summary, Darwin had to assume the production of a hereditary novelty for which there was neither evidence, nor a plausible mechanism, nor testable predictions, and so his theory did not have as rigorous and coherent a foundation as theories in the experimentally based sciences. However, the later recognition of genes, mutations, and recombination provided the required hereditary mechanism, and the development of population genetics made it possible to demonstrate microevolution occurring in nature. Moreover, extrapolation of macroevolution was supported by cellular and biochemical uniformities throughout the living world, as well as by the paleontological record (now very precisely dated) and by morphological homologies.

Today molecular genetics has provided a still firmer foundation for evolutionary biology. We now know in great detail many mechanisms that generate variation, and several of these facilitate evolution by providing larger steps than the 'point mutations' (single base changes) of classical genetics. In addition, molecular genetics has provided direct evidence, of two kinds, for macroevolution: divergences between species in DNA sequences parallel their divergences in morphology; and the contemporary occasional transfer of DNA between even very distant species reinforces their genetic continuity. These findings provide a novel, falsifiable foundation for establishing evolution as a process, quite apart from the history of its actual sequence and timing, inferred from the fossil record.

To be sure, we cannot predict in detail the future course of evolution — but that does not undermine its scientific solidity, despite Popper, any more than our inability to predict the weather far ahead should shake our confidence in the underlying physical principles. On the contrary, the position of evolution in biology is now comparable to that of Newton's laws in mechanics, or of atomic theory in chemistry. Accordingly, the term 'theory of evolution', often used to deny its solidity, is outdated. At the same time, the term 'evolutionary theory', like 'atomic theory', remains useful to designate a systematic effort to refine the modes of analysis and the concepts of the discipline.

In closing, we should recognize not only how much molecular genetics has contributed to evolutionary biology, but also how much it has unified the biological sciences. It has linked the formality of genetics with the concreteness of biochemistry; and along with electron microscopy it has opened up a forbidden range of dimensions, which long separated the visible objects of the microscopist from the invisible molecules of the chemist. Moreover, it has joined the two main streams of biology: the naturalists, asking how diverse organisms arose and are distributed, and the experimental physiologists, asking how they function.

Even more, molecular genetics has provided a rigorous answer to one of the deepest questions in biology: what fundamentally distinguishes living from nonliving matter? In vain, earlier vitalists sought novel forces, and some physicists more recently sought novel physical laws, such as Schrödinger's negative entropy. Now, however, we have a clear and marvelously simple answer. Biological systems employ only standard physical laws in the organization and function of their materials, but that organization generates a unique property: *the storage of information in molecular sequences.*

The concept of molecular information emerged from studies of the replication and expression of the genotypic program in nucleic acid. However, we should recognize that it is not limited to that material; it also applies to allosteric proteins, whose reversible changes in shape can detect information about their surroundings and can transfer it to responsive units in the cell. Moreover, while this property was originally discovered in regulators of gene and enzyme activity in bacteria, it also provides a molecular basis for sensory transduction in the recognition of stimuli in the environment; and the sophisticated networks of information transfer and storage in the nervous system of higher organisms are also based on allosteric effects, transmitted across cell membranes.

These developments have profound epistemological implications, for they illuminate the evolutionary and the functional basis for the continuity between our two kinds of knowledge: that programmed in our genes (which is *a priori* for individuals and *a posteriori* as a product of evolution) and what we acquire from the environment. Biology now tells us that Kant was right, and the empiricists wrong: we start our lives with a great deal of information, which is developed as well as supplemented by our subsequent contacts with the environment.

After focusing on such inspiring insights from modern biology, it is painful to return to the creationists. But they are there. Hopefully, emphasis on the elements of molecular genetics in the teaching of evolution not only should help a broad public to share gratification over the remarkable success of science in analyzing the complexity of living beings. It should also help the public to accept evolution as a reality, and to welcome its future contributions to our understanding of man.

Notes

1   Since Popper introduced 'falsifiable' as a technical term in philosophy, I use it here, though many scientists prefer the terms 'testable' or 'disprovable', and Popper has also used 'testable' and 'refutable'.
2   K.R. Popper, *The Poverty of Historicism* (London: Routledge and Kegan Paul, 1957).
3   K.R. Popper, *The Logic of Scientific Discovery* (London: Hutchinson, 1959).
4   P.S. Schilpp, *The Philosophy of Karl Popper* (LaSalle, Ill.: Open Court Press, 1974).
5   K.R. Popper, *Dialectica* 32 (1978): 344.
6   K.R. Popper, *New Scientist* 87 (1980): 611.
7   M. Ruse, "Karl Popper's Philosophy of Biology," *Phil. Sci.* 44 (1977): 638. Also in M. Ruse, *Is Science Sexist?* (Holland: Reidel Dordrecht, 1981), p. 65.
8   E. Mayr, *The Growth of Biological Thought* (Cambridge, Ma.: Harvard Universty Press, 1982).
9   W.B. Provine, *The Origins of Theoretical Population Genetics* (Chicago: University of Chicago Press, 1971).
10 E. Mayr, *Animal Species and Evolution* (Cambridge, Ma.: Harvard University Press, 1963).
11 T. Dobzhansky, *Genetics of the Evolutionary Process* (New York: Columbia University Press, 1970).
12 J.S. Huxley, *Evolution, the Modern Synthesis* (London: Allen and Unwin, 1942).
13 E. Mayr and W.B. Provine (eds.), *The Evolutionary Synthesis* (Cambridge, Ma.: Harvard University Press, 1980).
14 A.C. Wilson, S.S. Carlson, and T.J. White, "Biochemical Evolution," *Annu. Rev. Biochem.* 46: 573, 1977.
15 M.C. King and A.C. Wilson, "Evolution at Two Levels in Humans and Chimpanzees," *Science* 188 (1975): 107.
16 P. Zambryski, M. Holsters, K. Kruger, A. Depicker, J. Schell, M. Van Montagu, and H. Goodman, "Tumor DNA Structure in Plant Cells Transformed by *A. tumefaciens*," *Science* 209 (1980): 1385.
17 J.P. Martin, Jr. and I. Fridovich, "Evidence for a Natural Gene Transfer from the Ponyfish to its Bioluminescent Bacterial Symbiont *Photobacter leiognathi*," *J. Biol. Chem.* 256 (1981): 6080.
18 M. Busslinger, S. Rosioni, and M.L. Birnstiel, "An Unusual Evolutionary Behavior of a Sea Urchin Historic Gene Cluster," *EMBO J.* 1 (1982): 27.

On Experimental Approaches and Evolution
A Comment

Yadin Dudai

Two conceptual revolutions, which occurred a long time ago, shaped the face of modern research in the life sciences. The first was the mechanistic revolution, the cornerstones of which were formulated mainly by René Descartes in the seventeenth century. Animals, said Descartes, are machines governed by the same laws that govern any other physical object under the Sun. (Man, he added, has in addition a soul, that resides in the machine and interacts with it.) It is Descartes' mechanistic views that paved the way to the reductionistic approach characterizing most, if not all, of the work carried out today in biology laboratories. The second major revolution culminated two centuries later. The living world is not static, but undergoes continuous alterations and innovations; species were not created as such in the beginning of time but evolved from ancestral forms. Man is no exception. The clear-cut, poetic events of the third, fifth and sixth days of creation, as depicted in Genesis, were thus replaced by a seemingly cold and dry scientific alternative. For this revolution, the main responsibility lies with Charles Darwin — although he was not the first to initiate it.[1]

From a philosophical point of view, the mechanistic revolution was at least as important as the Darwinian one. But it seems that very rarely were laymen's (and probably also scientists') nights disturbed by thoughts on the implications of ultimate reduction as a strategy of analyzing living creatures.

111

E. Ullmann-Margalit (ed.), The Kaleidoscope of Science, 111–115.

In contrast, Darwin's ideas continue to haunt us persistently until this day. An English lady of the aristocracy, upon hearing Darwin's ideas for the first time, exclaimed: "I hope he is wrong, but if not, let us hope that the common people don't learn about it." The common people did, and most of them, with the aggressive back-up of clergymen and others who claim to know God's will, resisted the ideas. Evolution ceased to be a meta-theory confined to scientific circles; it became a public issue loaded with emotions. And so it remains today.

The strong emotions do not stem, so it seems, from the naive belief that we are all descendants of the baboon or the gorilla. Everybody is free to depict in his imagination the figure of his great-great-grandfather as he wishes, but the theory of evolution never claimed that we are the offspring of the apes that we keep locked up in our zoos. The theory does suggest that we and the apes shared, once upon a time, common ancestors. This indeed changes man's place in the universe, but by itself, I think, this piece of knowledge could have been swept somehow under the rug, had it not been for another popular claim: that evolution disengages us from God; on a superficial level, by discrediting a simplistic reading into the story of Genesis, and on a more profound level — by rejecting a teleological basis for our existence. The Darwinian revolution thus became an ideological revolution par excellence.[2] And an apparent deprivation of the biblical God is a strong enough blow to evoke extreme emotional debates, in which efforts are devoted to prove that evolution makes an interesting story, but not more than that.

Not that the meta-theory of evolution does not have its own share of scientific problems, but these problems, either major or minor, are often taken out of context by non-experts to gore the foundations of the theory. Professor Davis' thoughtful and expert exposition does not only deal with a major philosophical problem, related to the possible and available strength of experimental foundations of the theory of evolution, which should be of interest to every student of the life sciences; it also exposes the immense strength of novel scientific approaches and their ability to fill in gaps in the theory, both at the base of the pyramid as well as at the top. His lucid arguments should be carefully read by biology teachers: students should be enthusiastically encouraged to pose bold questions and expose flaws, and a pillar of science such as the theory of evolution should not, of course, be spared. But at the same time, proper knowledge and understanding of methods, theories and facts is a must. The last, almost trivial requirement is neglected by those who attack evolution today on theological bases, making use of misquoted bits of information. Professor Davis' paper should shed proper rational light on the subject.

I would like to add only a few, very brief comments, which might, I hope, prove to be of some interest with regard to our discussion.

How did it all begin? Our understanding of the possible mechanisms of evolution is now backed by several powerful scientific disciplines, and, as Davis has outlined, the novel tools and findings of molecular genetics provide direct evidence for the continuity of the whole biosphere, thus permitting extrapolation from microevolution to macroevolution. Science can thus cope with the mechanism of evolution in progress. Still, evolution must have had, in the remote past, a starting point. For biological evolution, it is the beginning of life. For cosmological evolution, it is the beginning of the universe. Science does not have a real solution for such singular points in which the process starts from zero. Cosmology offers the big bang, or the big bang that preceeded the big bang, etc., etc.; but what existed before that? Biology suggests vague mechanisms of *de novo* synthesis of reproductive molecules from nonreproductive molecules, but we would never be able to prove that life really started this way. Some eminent scientists, like G. Hoyle and F. Crick, apparently tried to evade part of the problem by suggesting indirect or direct panspermia, i.e. the importing of life from extraterrestrial sources. One of the reasons that led them to suggest this was the evidence for the existence of quite complex biological structures on Earth relatively shortly (on a geological time scale) after the formation of our planet. But removing the problem to space does not solve it. Thus we have powerful tools to demonstrate that evolution did happen and does happen; but we may stay forever ignorant about the most crucial point in cosmological evolution, or probably also in its biological subsystems: the starting point.

Evolution and beliefs. Debates on evolution formed an arena for intense clashes between science and anti-science, the latter usually being identified with religion. There are, in my opinion, elements of misunderstanding in definitions commonly employed in the debates. We often tend to draw excessive philosophical conclusions from our scientific findings, and sometimes even see them as substitutes for religion. The meta-theory of evolution is a scientific approach and hence can deal only with problems that fall within the realm of science (see above). It does not in any way negate or exclude the potential belief in a supreme power, in the most abstract meaning of the notion, for those who wish to believe; it contradicts only some elements of institutionalized religions and irritates their establishments. When Salvador Dali commented in 1964: "And now the announcement of Watson and Crick about DNA; this is for me the real proof of the existence of God"[3] — he merely proved that the interpretation of scientific data, when removed outside the realm of science, is in the eye of the beholder.

How important is evolution for us? Biological evolution shaped *Homo sapiens*, but the latter's brain created a novel domain, which is of course an extension of former evolutionary stages: cultural evolution, which is faster and more efficient than the organic processes that led to the emergence of the brain itself. We are very far from understanding the detailed rules of this new phase of evolution. Indeed, this cultural evolution enables us today to play at our will with basic tools of biological evolution, i.e. to practice genetic engineering. But at least at the present stage, this is a very risky method of shaping the future of our species. Culture and its artifacts are much more efficient. The evolution which is of great importance to our future is not a first-order product of DNA, since we are currently shaped mainly by higher-order products of our genes, via our cerebral hemispheres. The deciphering of evolutionary processes which operate on our brain and via our brain should be a major goal of our future research.

Ontogenesis does not yet recapitulate phylogenesis. The developments in molecular genetics have already granted us deep insight into complex mechanisms of genetic variations, that operate on the vertical (intraspecies) as well as on the horizontal (interspecies) axes. But it is worth emphasizing that, in spite of our much greater understanding of events that shaped phylogenesis, we are still very ignorant of the events that underlie ontogenesis. The rules and processes that transform the one-dimensional information coded in the genes into the three-dimensional embryo and adult, comprise a major component of the evolutionary process. They also comprise one of the greatest mysteries of modern biology and a major topic in the philosophy of the life sciences. It is very likely that further understanding of the mechanisms of gene expression and regulation will fill the void, and provide us with a new insight into the processes by which interaction of genetic and epigenetic factors unfold a living organism from a chemical code, and by which these rules of ontogenesis evolved themselves during phylogenesis.

The ever-existing surprises of science. Approximately 15 years ago, some prominent scientists were wont to explain that the golden age of the biological sciences is over, since the detailed mechanisms of inheritance are presumably known and, except for the brain, there is no remaining major frontier in the life sciences to be conquered. Not long afterwards, there came into existence the fascinating new field of genetic engineering. Not only did it disclose entirely unexpected facts about the mechanisms of transfer of our genetic material — it also shed new light on evolution, provided novel, direct molecular evidence for its dynamics, and in addition, enabled men, for the first time in evolution, to intervene rationally in the process itself. Thus one should never assume that the excitements of science are confined to *mémoires du temps perdu*. Our current ignorance of ontogenesis, and also of the mechanisms of mental operations, their relationship to the genetic material and their own evolution, only assure further excitement in the future.

Imagery and scientific revolutions. Great minds need not know too many details of a scene in order to draw a beautiful picture. Descartes, who hoisted the flag of the first major scientific revolution of modern times, claimed that half of his philosphy was completed when he came out of a stove, in which he spent an entire day to escape the cold of a Bavarian winter.[4] Darwin required more data and much more time, but he also apparently had a holistic image of a tree of evolution,[5] and formulated his ideas knowing nothing about the physical substrate that serves as the ultimate vehicle for the process, namely the genetic material in our cells. He built the frame and left it to us to modify parts and fill in the details at various levels of organization. Thus lack of detailed mechanisms should not hinder the productive imagination from wandering around and formulating working hypotheses. I sometimes toy with the idea that Darwin's grant application would not have been approved by some granting agencies had he lived today, due to the speculative nature of the project. But this relates to a particular corner of evolution: the evolution of our scientific establishment and its paradigms. And that is already another story.

Notes

1   E. Mayr, *The Growth of Biological Thought: Diversity, Evolution and Inheritance* (Cambridge, Mass.: Harvard University Press, 1982).
2   K. Popper, "The Rationality of Scientific Revolutions," in: *Scientific Revolutions,* ed. I. Hacking (Oxford: Oxford University Press, 1981).
3   G. Stent, *Paradoxes of Progress* (San Francisco: W.H. Freeman, 1978).
4   B. Russell, *History of Western Philosophy* (London: George Allen and Unwin, 1961).
5   H. Gruber, "Darwin's 'Tree of Nature' and Other Images of Wide Scope," in: *On Aesthetics in Science,* ed. J. Wechsler (Cambridge, Mass.: The MIT Press, 1981).

Darwin's Principle of Divergence as Internal Dialogue

DAVID KOHN

Introduction

However strongly we may see scientific ideas as socially and culturally contingent in their origin and expression, we must acknowledge that they are also the products of individuals. Hence, even if we consider all scientific activity to be but a reworking of prior scientific activity, the dynamics by which individual scientists develop their theories is a subject integral to the history of science. If we accept the proposition that knowledge grows by public and critical dialogue, we should not ignore the fact that important phases of the dialogue may occur within an individual. Such is the case with Charles Darwin, who over the decades prior to the publication of *Origin of Species*, engaged in an extended reworking not only of natural history, but also of his own emerging ideas. For a scientist such as Darwin, the internal personal debate is as fierce and as fertile as many a public debate.

The subject of my paper is the internal dialogue that produced Darwin's principle of divergence.

The principle of divergence was in a very real sense a culmination of the most important stream of Darwin's evolutionary thought. It was only fully formulated in the mid-1850s, although there are many important anticipations earlier, as Sam Schweber has shown. In effect it was the last

117

E. Ullmann-Margalit (ed.), The Kaleidoscope of Science, 117–132.
© *1986 by D. Reidel Publishing Company.*

crucial intellectual step that Darwin took before writing the *Origin*. The major focus of my paper will be a close examination of how Darwin took that last step. My historiographic microscope will be ratcheted down to high power. I am doing this out of the conviction that only when we take a very close look at the texture of thought of a scientist, can we grasp the precise way in which the scientist operates and embodies the larger scientific trends of culture. In other words, I am proceeding from the conviction that there exists a need not only to do case studies but to do very close exegetical case studies. There is a need to do this because otherwise, when we come to do higher order comparative studies within a period, we risk constructing some very misaligned bridges between the so-called external and internal levels of analysis. I am reminded of a wonderful advertisement I used to see in England for the *Sunday Times* showing two roadbeds for a bridge built from opposite directions, and of course the spans were meant to meet in the middle. Unfortunately, however they were misaligned by about six feet. The caption for this ad was: Don't you wish you were better informed? So that is the spirit in which I do my work.

Now before getting down to my exegesis, let me briefly sketch out the chronological background of Darwin's intellectual development.

In late 1836 Darwin returned from the five years' voyage of the *Beagle*. The majority view among Darwinists is that he became firmly convinced of evolution only in England, in the first quarter of 1837. Sulloway has shown that Darwin first arrived at this conviction when the ornithologist John Gould informed him that certain Galapagos birds collected by the *Beagle* expedition were species distinct from those found on the mainland of South America. It was the recognition that these were new species, rather than mere varieties of known ones, that led him to evolution. Once he accepted evolution as the framework of biological reality, he never deviated from that position, and the problem became the derivation of an explanatory theory of evolution. The principal locus of his search for an explanatory theory to work by is in his transmutation notebooks maintained from July 1837 through 1839. Numerous scholars have shed light on this search. In my own reconstruction I have stressed that over this period Darwin formulated, and later rejected or modified, a *series* of coherent theories. At each stage the prevailing coherent theory constrained the way in which he evaluated the data of natural history. In September 1838 this process culminated, through Darwin's reading of Malthus on population, in the formulation of natural selection as the key to explaining evolution. First in 1842 and again in 1844 Darwin wrote unpublished essays that presented his theory and applied it to the major branches of natural history. The striking feature of these essays is

Darwin's stress on the limited amount of hereditary variation found in nature. Since natural selection depends on such variation, we find Darwin granting a far more limited scope to the power of natural selection than the mature Darwin familiar from the *Origin*.

From 1847 to 1854 Darwin became obsessively preoccupied with his taxonomic work on the barnacles. One should see these seven years not as a dead period but rather as a period of latency during which he continued to read and speculate and to amass a considerable volume of loose notes. In September 1854 he packed up the last of his barnacles and began to devote all his energies to the species problem. In 1856 he began writing the long version of the *Origin*, published posthumously as *Natural Selection*. In June of 1858 his work was interrupted by the famous letter from Alfred Russel Wallace. Shortly thereafter Darwin set himself to writing an abstract of the long version, which of course was published in November 1859 as *Origin of Species*. The significant theoretical addition to Darwin's thought in the post-barnacle period was the principle of divergence which, Darwin said, was "together with natural selection a keystone of my book." So much for a potted sketch of Darwin's intellectual development from 1836 to 1859.

Let me now proceed to define the principle of divergence. The argument Darwin called the principle of divergence runs as follows:

1. First there is an ecological premise. A locality can support more life if occupied by diverse forms partitioning resources. This is the ecological division of labor. Thus specialization is an adaptive advantage to an organism. Hence natural selection, which explains the origin of all adaptations, favors the evolution of new specialized varieties.

2. The emergence of a new variety occurs sympatrically, that is, with parental and offspring forms inhabiting the same locale. Thus the making of varieties, which Darwin saw as incipient species, occurs by vigorous selection for specialization that overcomes the swamping effect of crossing.

3. From this first fork of the branching phylogeny it is a matter of reiteration to generate all of classification. Simply put, niche within niche engenders group within group.

Darwin's principle is itself a set of nested arguments comprising the idea of natural selection, the idea of speciation without isolation, and the view that the relations among organisms create new evolutionary situations. One thing about the argument stands out: it is internally unified by natural selection. That is, the explanations at the three classic levels of evolutionary process — adaptation, species formation, and the hierarchical classification of organic diversity — are portrayed as the application and consequence of natural selection.

One difficulty in studying Darwin one hundred years later is that we all are, or believe we are, Darwinians. Modern evolutionary theory — including the postsynthesis versions — comes equipped with its own historiography, which includes a view of Darwin. In particular, Darwin's theory of natural selection has been reduced to an explanation of one aspect of the evolutionary process: the origin of adaptation. For Darwin, however, selection explained more than adaptation. It was always intimately bound up in his thought with the multiplication of new species. The dominant modern explanations of species formation do not attribute to natural selection in this process the same role that Darwin did. The question is complicated by the oft-repeated slogan of the modern theory that evolution is the unifying theory of biology. This slogan rests on a number of claims, one of which is that evolution explains a hierarchy ranging from adaptation, through species formation, to the major trends of organic diversity. However, modern explanations of these three levels are not continuous, i.e. they do not employ the same mode of argument to explain processes at each level. Thus the modern theory, including the postsynthesis versions, is not internally unified. It is a chain of explanations, each more or less appropriate to its own level. This disunification may be one of the keys to its success. Nevertheless the slogan remains and some of its other claims may be justified, as Beatty's work suggests. Like his modern "followers," Darwin saw evolution as a unifying theory; unlike them he sought a theory that was internally unified by natural selection, and he thought that he had accomplished this goal through the principle of divergence. One would assume that there ought to be a basic disparity in the way in which a historian and a biologist would interpret Darwin's principle of divergence. For the historian Darwin's principle is part of his thought. It requires explanation in context. For the biologist the principle is something that Darwin got wrong. Since the biologists have, at least in the past, maintained a claim to be Darwinians, they have either wanted to sweep Darwin's error under the rug or to figure out where Darwin made his mistake. This has had its impact on historians. We have tended to follow the biologists' lead and have either ignored divergence or considered it a curiosity.

The past decade has witnessed important efforts, first by Limoges and then almost simultaneously by Browne, Ospovat, Schweber, and Sulloway, to reassess the principle of divergence. These have been stimulated by two factors inherent in the recent history of Darwin studies. First of all, our period has witnessed a considerable concentration on Darwin's intellectual development. To get beneath the often enigmatic surface of the Darwin of the *Origin*, the present generation of Darwin scholars has been led to study how Darwin's ideas were formulated. Inevitably, the course of this work followed the course of Darwin's life. Workers have tried to understand the process of his conversion to transformism, his first formulations of evolutionary explanation, and of course the construction of natural selection. By around 1977, the time had come to tackle the period from 1844 leading up to the writing of the *Origin*. For that phase of Darwin's career the principle of divergence stood out as an intellectual development. It was an idea whose turn to be studied had come. The second factor directing attention to divergence was access to the necessary materials. The publication of *Natural Selection* by Professor Stauffer in 1975 was the primary stimulus. This led scholars, under the guiding hand of Sydney Smith, to attempt to make sense of the great collection of loose notes Darwin had accumulated in the 1840s and 1850s. In these notes, which Darwin organized in topical portfolios, was thought to lie the evidence for reconstructing Darwin's formulation of the principle of divergence. The contributions of Browne and Ospovat, and the one presented here, particularly derive from their access to both Stauffer's edition of *Natural Selection* and from Smith's efforts to organize and date the portfolio notes.

Suddenly we have had a great deal of light shed on the principle of divergence. Yet I feel more clarification is needed. We need to have a better understanding of two things: why we are studying the principle of divergence and how to go about it. My remarks so far suggest an answer to the first issue. In the construction of their theories scientists not only seek to explain natural phenomena. They often also have methodological goals. The search for unity is one such goal. As historians we ought to be cautious neither to worship nor to ignore the powerful lure of unification. We ought to recognize it as a recurrent tendency in scientific debate, that is often internalized as a value in individual scientific practice. In other words, it can guide the content of science in the same way that ideology has long been known to do. This was the case for Darwin's principle of divergence. He conceived of it as a principle of unification. This is why he called it "a keystone of my book." Unification was not a detached value. It was an internal guide that penetrated and directed the development of his research. This is the principal message of the story I am about to tell. The second matter that needs clarification is methodological. The progress achieved since 1959 in understanding Darwin's development came when we identified and clearly analyzed concrete episodes in Darwin's career. This is true for Darwin's conversion, his first theory, and natural selection. The principle of divergence, however, had a much longer period of gestation than those episodes. We still need both a descriptive and a causal embryology for the development of the principle of divergence. I propose now to define and analyze one episode in that developement.

Defining an Episode

Given the importance of analyzing *episodes* in a developmental approach, it occurred to me that a primary step was to identify some episodes. Let me give you an impression of the archival problem. We are dealing with several hundred loose pieces of paper distributed, not in the order in which they were written, but in topical portfolios. Furthermore, within each portfolio there is no reason to assume that the notes reflect Darwin's writing order. Given this lack of stratigraphy (a problem that does not exist with Darwin's notebooks), it seemed to me that the safest course was to concentrate on those notes that Darwin took the trouble to date. For the period 1852 through 1857 this came to 124 notes. I was somewhat reassured in this approach when I found few undated notes that I felt troubled about ignoring. When rearranging the dated notes from all the portfolios in chronological order, I had two dominant impressions. First of all I was impressed, and to a degree depressed, by the low density of dated note-taking. During the 72 months of 1852–1857 there are 33 months with no dated notes at all, 16 months with 1 note per month, and 15 months with 2–4 notes. Thus during over 80% of this period, note-taking, at least dated note-taking, can only be characterized as a sporadic activity. My second impression was that one date dominated all others in frequency: November 1854. There are 19 notes so marked. Thus some 15% of the dated notes from the six years of interest were written in that one month of 1854. At last I had at least one episode around which to build a theory. Lest the reconstruction of a period in Darwin's thinking from 19 scraps of paper appear an implausible object of study, think of the cogent theory, and indeed program of research, that Darwin concentrated into the first 20 or so pages of the B notebook.

November 1854 in Perspective

The November 1854 episode came at a point in Darwin's career that one would predict to be of interest. In September 1854 Darwin took a critical step toward the eventual writing of the *Origin*. He sorted the loose notes he had accumulated since writing the 1844 Essay and distributed them into topical portfolios on biogeography, classification, hybridity, variation, embryology, paleontology, and behavior. In November 1854, having no doubt completed a review of his old notes, he, in effect, initiated the process of writing his species book by writing a spate of new notes.

Before looking at the details of Darwin's position in November 1854, it is well to put the conceptual boundaries of this episode into perspective. It is evident that Darwin had already set himself the task of showing that divergence was a tendency in nature that required a mechanism or an explanatory principle. Ospovat showed that, contrary to his position in 1844, when Darwin rather took divergence for granted, by 1847 he was already inclined to see it as a problem to be solved. Ospovat explained this shift as Darwin's response to those comparative anatomists, such as Milne-Edwards following on Von Baer, who held a branching conception of systematic relationship, but who maintained a creationist outlook. I think it might be profitable to see Darwin's shift as part of his more general effort to translate the several theories of contemporary systematics into evolutionary terms. This included an evolutionary reinterpretation of Swainson, Owen, and Strickland as well as of Milne-Edwards and Carpenter. But this is an area for future study. At any rate, the problem of explaining divergence was already constitutive to Darwin's thinking prior to November 1854. In contrast, the nub of Darwin's solution to the problem of divergence, namely ecological division of labor, was formulated shortly after November 1854. So this episode is the original locus of neither the problem nor its solution. Nevertheless, I see it as the turning point of the story, first because here Darwin consolidated his argument and established its characteristic unified structure, and second in structuring his argument he discovered here the particular line of reasoning that led subsequently, and rather quickly, to the division of labor.

Turning now to the core of the episode, we find that the focus of Darwin's attention was the use of biogeographic data to draw conclusions on the pattern of divergence. As Janet Browne has shown us, he worked in the botanical arithmetic tradition of Humboldt, Robert Brown, and A.P. de Candolle. In this tradition the evidence of geographic distribution was tabulated and summarized in ratios, typically expressing the number of species per genus in a particular geographic area. By comparing such ratios, general conclusions were drawn about patterns of distribution. In Darwin's hands this method became a powerful tool to show that the contemporary data of present distribution patterns could be used for two related ends: (1) to portray the stages in the historical process of evolutionary divergence; and (2) to portray the hierarchical classification of natural groups as the product of evolutionary divergence.

In November 1854 Darwin in fact began two processes. The first was the actual tabulation of data from catalogues, monographs, and synoptic works such as the Candollean *Prodromus*. Tabulation became a major project that continued into 1858. The other process was the conceptual one of drawing conclusions from tabulated data and establishing hypotheses to be tested against tabulated data. Out of these two processes the solution to the problem of divergence was formulated. According to Browne and Ospovat, the twin processes of tabulation and conceptualization went on hand in hand from 1854 to 1858. Thus they see Darwin substantially modifying his views on the bearing of his data on the principle of divergence as he made new calculations. This led Ospovat to characterize the formulation of the principle as occurring gradually over a period of years, and Browne to see it as happening in 1857, near the end of the calculations. Unfortunately, I have to disagree with the basic premise of both my friends Browne and Ospovat. As I see it, in November 1854, when Darwin made his first calculations which he considered to be a success, he became convinced that the botanical arithmetic approach would allow him to prove certain hypotheses.

Starting from this premise, my thesis is threefold: 1. That Darwin drew his complete set of conclusions from the biogeography of living groups in November 1854 and simply tested these by laborious calculations over the succeeding years. The long series of calculations always confirmed Darwin's hypotheses. In other words, this episode is one more instance of the priority of theory over evidence in Darwin's intellectual development. 2. That also in November 1854, he used a particular form of historical reasoning to transform his biogeographic conclusions into a proof that the history of life was divergent. 3. Finally, that in November 1854 his transformation of biogeographic data into a historical narrative established the framework from which he drew the further critical conclusions necessary to complete his principle of divergence. The first of these conclusions was that speciation can occur by pure sympatry, that is, without any form of prior isolation. This was a temporary position from which he retreated. However, it provided the intellectual emancipation that led to the second critical conclusion, namely that species are formed not just by natural selection, as he had long believed, but that species multiply in those ecological conditions that permit vigorous selection. In sum my thesis means that the central structure of Darwin's argument for divergence, with its characteristic unification of natural classification, speciation, and ecology, and the defining of conditions for the ecological division of labor, were all dashed off (as it were standing on one foot) in November 1854. The division of labor itself, which completed the principle, was added three months later, in January 1855. All the rest, including writing the divergence sections of *Natural Selection*, was calculation, revision and exposition.

This is my thesis; let me see if I can put some evidentiary flesh on these bare bones.

Biogeography as Historical Narrative

Although Darwin's calculations over 1854–1858 were massive, their basic logic was in place already in November 1854 and they can be succinctly summarized. He made four basic calculations; two were focused on small genera and two on large genera. The starting point was small genera. In November 1854, he had George Robert Waterhouse mark the aberrant or peculiar genera in Schoenherr's catalogue of the *Curculionidae* (the weevils). He calculated the number of species per aberrant genus, and found that the aberrant genera had fewer species per genus than the average number of species per genus in the family as a whole. Second, he also examined the geographic range of all small genera in the curculionids, be they aberrant or not. He found that the aberrant genera had scattered ranges, and the non-aberrant small genera had predominantly local distributions. Thus, Darwin thought he had discriminated between two kinds of small genus. He interpreted small genera with local distributions, where the species are morphologically closely allied, as rising or nascent genera. On a sheet summarizing his calculations from Schoenherr he wrote:

All rising genera must be local <(& closely allied>: ...

In contrast, small genera with morphologically very distinct species — distinct enough to be called peculiar or aberrant — were found to have a scattered distribution. He interpreted these aberrant genera, as he had since the 1830s, as dying genera, or living fossils. To continue reading from the same note:

... all dying genera, <with species very distinct> ... wd be small, aberrant & <if they had died equally over world, wd be> widely distributed ...

In other words, the biogeographic data could be used to identify nascent genera by their pattern of species fanning out or diverging in the local area of their birth. It could also be used to characterize the end-point of divergence as the scattered and peculiar relics of what might once have been large genera.

These then were the hypotheses Darwin tested and the evolutionary inferences he drew from the work on small genera in the curculionids.

However, in November 1854 Darwin also saw that the link between small nascent and small dying genera was formed by large genera. He wrote:

> ... if extinction has fallen near & around the aberrant genera, then *creation* has fallen on the typical & larger genera. —

In November 1854 Darwin saw two hypotheses to be tested with respect to large genera: (a) that the species in large genera tend to be wide ranging; and (b) that the species in large genera tend to be polymorphic, that is, broken up into varieties. Concerning range, he wrote:

> Undoubtedly larger genera are large partly because they are widely distributed & have representative species in different countries.

This hypothesis has a very definite history in Darwin's thought. From the 1830s Darwin had sought to make some sort of statement about worldwide genera and species. He produced a number of indefinite and contradictory statements until 1845, that is, after the 1844 Essay was written, when he found in Swainson the proposition that by "typical genera" systematists meant large genera, and that large genera were wide ranging. There is a rich correspondence between Darwin and Hooker in 1845 that reflects Darwin's attempts to have these propositions confirmed by Waterhouse and Westwood. From that time the relationship between wide range and large genus became, I think, something of an *idée fixe* that Darwin turned over in his mind.

Hand-in-hand with wide range, went polymorphism. He wrote:

> To explain why the species of a large ≪(& consequently polymorphous≫ genus, will hereafter probably be a Family with several genera, we must consider, that the species are widely spread & therefore exposed to many conditions & several aggregations of species: ...

Implicit in this proposition is Darwin's long-held view that varieties are incipient species. Darwin began the calculations to test these hypotheses in January 1855. Using Hooker's *Flora of New Zealand*, he calculated the number of species that present varieties, and he found more such polymorphic species in large genera than in small genera. He applied calculations of this type to an ever-expanding number of botanical works through early 1857. From August 1855 he began calculating the number of species with wide geographic ranges, and found more wide-ranging species in large genera than in small ones. Again he applied this calculation to many botanical works through early 1857. Throughout this undertaking Darwin's calculations confirmed his expectations. In July 1857, John Lubbock informed him that his calculations were in error because he had not precisely defined small and large genera. He then repeated all of his laborious calculations on large genera, which took him well into 1858. He concluded that the new calculations also confirmed his expectations.

I hope that the first point of my threefold thesis is evident from the foregoing: Darwin's calculations only confirmed the hypotheses he held in November 1854. Furthermore, most of these hypotheses were long-held constructs that had implications for divergence. Which brings me to my second point. Out of this well-seasoned timber Darwin built the following description of divergence:

> Hence, <small,> genera will be local <owing to> — their origin <from common parent>; & small genera (...) certainly, from Schoenherr, are local in proportion of 215:52 (...). As to make species is slow work, & [to make] genus increase to <considerable> size much time would be required, hence as Forbes says [they] wd be local in their origin in past time; the species wd extend over continuous spaces in area & time. But it wd generally happen during the time necessary to make a large genus, that geographical mutations & chance accident wd disperse genera & then [the] very fact of the genus having become large in one area, we may suppose wd give it some better chance in another area, & thus the genus wd get bigger & bigger, and certainly most large genera are widely extended. When a genus began to fail & die out, if large, it wd leave probably a few species in distant quarters of the world: Hence this would be another cause of small genera: these wd be aberrant [.]

It is clear here that Darwin has taken biogeographical patterns, demonstrable in the present, to represent the historical stages of divergence. Biogeography supplied what the inherently imperfect fossil record could not: a coherent historical narrative. Stephen J. Gould has recently observed that the argument in most of Darwin's so-called minor books is covertly structured by historical reasoning. He views these lessons in historical reasoning as among Darwin's most lasting contributions. He also recognizes three categories of historical reasoning in Darwin, which are distinguished by the amount and nature of available evidence. Gould describes one form of historical reasoning as follows:

> If rates are too slow or scales too broad for direct observation, then try to render the range of present results as stages of a single historical process.

This, I think, is exactly what Darwin did in November 1854. He transformed biogeography into a historical science to 'prove' divergence. Moreover, there is evidence that Darwin was methodologically self-aware that he had found a way to read the past from the present. He commented:

> I am inclined to think that it is very curious how similar all laws of relations between organisms separated by time & space;...

And on one of his undated slips he noted:

> Space & time analogous.

This methodological self-awareness in 1854 strongly echoes the early passage in the first transmutation notebook where Darwin wrote:

> ... as we see them in space, so might they in time ...

As Howard Gruber observed, it was shortly after penning this remark that Darwin sketched his first branching diagrams of the tree and the coral of life. What separated Darwin of July 1837 from Darwin of November 1854 is that he had identified and hoped to quantify those particular patterns in space that showed what species might become in time. It was his specific bigeographic hypotheses, reworked over years and clarified in dialogue with his self-chosen colleagues, that gave substance to the historical analogy between space and time.

But the goal of Darwin's biogeographic work in November 1854 was not merely to show that the history of life was divergent, it was to show that this history accounted for the hierarchical, hence tree-like, natural classification. His goal was, as he wrote:

> ... [to] explain why the species of a large genus will hereafter probably be a Family with several genera ...

It was to show that the theory of descent

> ... give[s] the diverging the tree-like appearance to the natural genealogy of the organised world.

It was to show that natural classification is a "natural genealogy."

Speciation and Natural Selection

What further distinguishes the November 1854 episode is Darwin's determination to find an explanatory mechanism for divergence. This brings me to the third and final point of my thesis. On the paper containing the Schoenherr calculations, Darwin wrote a long note under the title "Theoretical Geographic Distribution." Here he addressed the problem of speciation. As I have shown elsewhere, Darwin in July 1837 elaborated two models of species formation: a phyletic model appropriate to a continuous range; and a geographic-isolation model appropriate to islands. The phyletic one came close to sympatric speciation inasmuch as Darwin recognized that this model would be bedevilled by blending inheritance. As Sulloway has shown, Darwin came to strongly favor the geographic-isolation model into the early 1850s.

However, in November 1854 Darwin took a fateful step. He wrote:

> When a species breaks & gives rise to another species, the chances seem favourable (...) to its giving birth to others. [No doubt here comes in question of how far isolation is necessary, & I shd have thought more necessary than facts seem to show it is] In fact there never can be isolation for the parent forms must always be present & tend to cross & bring back, to ancestral form; it will *always* be a struggle against crossing, & will require either vigorous selection or some isolation from habit, fewness [,] nature of country to separate. Hence, small genera will be local <owing to> their origin <from common parent>; ...

Here we have Darwin firmly turning his back on the strict necessity for isolation in favor of pure sympatric speciation. We can feel the tense strain of this movement as he wrote: "I should have thought [isolation] more necessary than facts seem to show it is." In fact he establishes a dichotomous cleft. Species may be multiplied *either* by vigorous selection, with no isolation, or by various indirect but effective isolating barriers: behavioral shifts, low population density, partial topographical barriers. I will have more, presently, to say about this dichotomy, to which I attach considerable importance. But our attention should now be focused on the implications of pure sympatric speciation, where intense selection is seen to be as powerful a force in breaking species as a mountain chain or an ocean. It is clear from the text that the immediate factor behind Darwin's shift away from isolation is his biogeographic work on the proliferating species of small nascent genera. As he came to see the species of local genera as the primary locus of divergence, he came to regard small locales with no chance of geographic isolation as the primary sites of speciation. Appreciating full well the swamping effect of crossing, he is forced to invoke vigorous selection as the only effective countervailing force. But, more important, this line of thinking leads Darwin to look for the local, ergo ecological, conditions that favor vigorous selection. The focus of his attentioin goes to the biotic interactions of assemblages of organisms in small and uniform areas. He writes:

> It is indispensable to show that in small & uniform areas there are many Families & genera. For otherwise we cannot show that there is a tendency to diverge (if it may be so expressed)...

In other words, his attention goes to the ecology of crowding. It is there that he expects to find the reason for the "tendency to diverge."

This is as far as Darwin went in November 1854. The characteristic three-tiered structure of the principle of divergence is in place. Biogeography allowed him to reconstruct the history of life as divergent, which allowed him to construe branching natural classification as a consequence of divergence. That is the first tier. His biogeographic work on local genera focused his attention on speciation in a locale without isolation. The result was sympatric speciation by vigorous natural selection. That is the second tier, which led him to look directly for the ecological conditions where vigorous selection would prevail, namely: crowded small and uniform areas. That is the third tier. He has conceived an integrated structure that is unified by natural selection. He has yet to complete the structure by the ecological division of labor. But he knows what he wants and he knows where to look for it.

Division of Labor

The missing element that breathes life into this structure is found in a note dated 30 Jan 1855:

> On theory of Descent, a *divergence* is implied & I think diversity of structures supporting more life is thus implied. ... I have been led to this by looking at a heath Thickly clothed by heath, & a fertile meadow, both crowded, yet one cannot doubt more life supported in <second> than in first; & hence more animals are supported. This is not final cause, but more [a] result from struggle, (I must think out this last proposition) —

The idea here is very simple: as a result of struggle more life can be supported in a meadow with its diverse flora than in the monoculture of a heath. He does not call this idea the ecological division of labor. Instead this is the idea that he later compared to the division of labor. The label was not applied until September 1856. To mistake the labeling for the conception would, I believe, be a misinterpretation of Darwin's developmental process. In my view the principle of divergence was structured in November 1854, including the form of the solution, and by or before January 1855 Darwin had his "keystone."

Conclusion

On this reading the switch to sympatric speciation was a watershed. I will conclude by returning to Darwin's dichotomous views on speciation. We saw that as an alternative to what became species formation by the principle of divergence, Darwin recognized various forms of isolation. In *Natural Selection*, indeed in the section on divergence, Darwin discusses the conditions for speciation and he opposes natural selection to isolation. But it seems to me he has convinced himself that, as Sulloway concluded, some form of isolation is almost always necessary. He adds complete and partial geographic isolation to the ethological and habitat barriers he recognized in November 1854. He also diagnoses the degrees of isolation required according to the breeding system and mobility of the organism. Animals that are highly mobile and freely crossing require the most intense isolation. Plants that do not cross for each birth and are sessile but may hold ground by proliferating rapidly require less isolation, but, of course their breeding system and habit are kinds of isolation. But even in this case he says:

> I can well believe that a selected variety might be more quickly formed & hold their own against the ill effects of crossing, without being completely isolated. Though in such cases, isolation, at least partial isolation at first, would be favourable to their natural selection.

The simplest way of putting the situation is to say that Darwin wanted to have his cake and eat it. He wanted speciation by natural selection alone, but he was in fact a rather woolly isolationist. The reason why he never resolved this internal contradiction is plain. The principle of divergence, which he valued for the unification it gave to his theory, was grounded both conceptually and, perhaps more important, developmentally in sympatric multiplication of species by vigorous selection.

I believe it was F. Scott Fitzgerald who said it is the mark of a great artist to be able to hold two mutually contradictory ideas simultaneously. Darwin was a great scientist. But his internal contradiction certainly confused his followers. Mapping the conflict in the reception of Darwin's theory, at least among those English naturalists who considered themselves Darwinians, we find that the lines of demarcation follow the internal lines-of-cleavage formed during the development of Darwin's thought. The late nineteenth century found Romanes and Gulick pitted against Wallace over opposing resolutions of Darwin's contradiction, with Wallace championing selectionism and Romanes and Gulick laying the ground for isolationism, the dominant modern view. Ultimately, the issues I have been discussing suggest that the structure of post-Darwinian debate reflects the dichotomous structure of Darwin's thinking. There is no intention here to canonize Darwin. Just as in a Moebius strip there is only one side, so in the history of science there is only reception within scientific communities. But to understand *that* public critical dialogue through which knowledge grows, the case of Darwin's principle of divergence shows we must attend, with careful scrutiny, to that internal dialogue which is individual intellectual development.

On Darwin's Principle of Divergence
A Comment

SILVAN S. SCHWEBER

In a valuable article on "the theories to work by" that Darwin formulated from 1835 to 1838 to account for the origination and extinction of species in nature,[1] David Kohn gave an impressive panorama of Darwin's intellectual development during that period. This work, which paid meticulous attention to Darwin's scientific theorizing, is the most accurate and reliable account of Darwin's views regarding adaptation, variations, and speciation before coming to Malthus, and shed new light on the Malthusian impact.

Dr. Kohn has maintained the high standards he set in that work and has given us an equally impressive account of Darwin's theorizing in the period from September 1854 to the spring of 1855, when he was getting started on writing his Big Species Book. The work this time is more narrowly focused: He has done so to emphasize the crucial importance that Darwin attached to his principle of the divergence of character in the construction of the *Origin* (or rather in the Big Species Book, *Natural Selection*, the treatise Darwin was writing at the time). It is a fascinating and important piece of work which clarifies the structure of the argument by revealing its genesis. It is Darwinian in that it explains by giving the history.

What Dr. Kohn hasn't done, or has not yet done but which can now be done, is to compare the style and the approach of the Darwin of the 1835–8 period to the more mature and accomplished Darwin of the fall of 1854 and

133

E. Ullmann-Margalit (ed.), The Kaleidoscope of Science, 133–136.
© 1986 by D. Reidel Publishing Company.

the spring of 1855. In both periods we are looking at "internal dialogue," but the personae holding the dialogue are vastly different.

The theorizing of the young Darwin in the 1835–8 period is totally private, essentially secretive, with very little empirical activities going on simultaneously. His transmutationist, naturalistic explanation of the origin of new species (including man) was heretical and his theory would surely have caused his ostracism by his scientific community and very probably also by his social community, had he published it. It is likely that Emma Wedgwood would not have married him had he divulged to her his views on man's place in nature. The Darwin of 1838 was regarded as an upcoming geologist, not yet of elite status, and an accomplished collector and natural historian.

The Darwin of 1854, on the other hand, was recognized as a first-rate systematist, paleontologist, and zoologist — having just completed his extensive work on the cirripedes — and as an outstanding geologist who had stopped working in that field some ten years before. He was widely known to hold heterodox views on the species question. He had divulged his species theorizing to Hooker, and more selected aspects of it to others, e.g. Waterhouse. He was a respected member of his community at Downe. Evolution by that time was no longer heresy. Another aspect of Darwin's theorizing of that period should be stressed and not overlooked. Intimately related to the theorizing Darwin embarked on in September of 1854 were the extensive experimental activities that he undertook at the same time. The centrality of these experiments to the writing of the Big Species Book cannot be overemphasized — and they are very much part of the internal dialogue. Experiments on the hardiness of seeds, on the relation of fertility of crossing in and between wild and domestic plants, on the embyrological development of different races of pigeons, on the yield of pastures as a function of the genera and families planted, etc. were the empirical foundations upon which his various hypotheses rested. Their result critically determined the nature of the "long argument." There is another aspect of Darwin's 1854–6 theorizing on botanical arithmetic that I believe ought to be stressed. It verifies the growing importance of statistics in the first half of the nineteenth century and illustrates the growing acceptance of *statistical* laws as explaining phenomena, or conversely of the use of statistical regularities as the basis of theoretical structures. The statistical nature of the data was thought to be the result of the complexity of the phenomena investigated, but, more important, the data were taken to represent *real* effects in nature — i.e. the scatter was not assumed to be errors from which an ideal, typical, true measure could be extracted.

I want next to turn to Dr. Kohn's paper and raise a substantive question. I am convinced of the centrality of the arithmetical botany and its use in putting the argument together. I can only convey my own enlightenment and pleasure in reading his all-to-concise and compact paper and being shown how Darwin used historical reasoning to transform his biographical conclusions into a proof that the history of life was divergent.

Kohn's emphasis on the importance to Darwin of the possibility of sympatric speciation — not only as a theoretical concept but also, significantly, as a psychological gestalt switch — is surely a very important observation.

That Darwin had the "keystone" of the argument by January 1855 is probably correct, but I would also suggest that the argument was still not complete in an important way — at least insofar as an explicit presentation is concerned. All the arguments up to that point referred to levels of descriptions above individuals: varieties, species, and higher taxa. Natural selection operated on individuals, and the linkage by which diversity is accomplished had to be explicitly stated. Darwin had obtained that insight from his earlier reading in political economy and agricultural chemistry.[2]

The principle of divergence in the form expressed in the Asa Gray letter of September 1857 is explicitly in place only after September 23, 1856, when Darwin recorded in a note to himself that:

> The advantage in each group becoming as different as possible, may be compared to the fact that by division of labour most people can be supported in each country. — Not only do the individuals in each group strive against the other, but each group itself with all its members, some more numerous, some less, are struggling against all other groups, as indeed follows from each individual struggling.

There is no question in my mind that the insight expressed in this note was part of the theoretical baggage that Darwin had carried for many years, and was a tacit component in the previous argumentation. The interesting question is why did Darwin, when referring to it in *Natural Selection* and in the *Origin*, call it the physiological division of labour and attributes it to Milne-Edwards, the zoologist, rather than to the political economists who, from Adam Smith to Darwin's time, had made division of labour the central concept in understanding the growth of the economy.

Finally, having read David Kohn's paper I am again struck by the appropriateness of the insights that Gruber has given us on the creative process. Gruber could just as well have used Darwin coming to divergence of character to illustrate his position vis-à-vis Aha Experiences, instead of

Darwin coming to Natural Selection on September 28, 1838. Let me quote Gruber, essentially verbatim, inserting only such changes as to make his remarks apply to the present situation.

> [Darwin's] thinking could be described as a process of purposeful growth organized into a number of distinct enterprises. These enterpises moved forward more or less in parallel. Within each he had many insights. To be sure, he was looking for a way of synthesizing these efforts, and he did indeed.

And here is where a change from Gruber's text has to be made: One day in January 1855 — or, if you prefer, during the famous carriage ride to which Darwin alludes in his autobiography, though we don't know when it occurred — he did have a major insight in which he saw clearly how eventually to formulate the principle of divergence of character. This is the analogue of the September 28, 1838, recollection of Darwin that he read Malthus "for amusement." And returning to Gruber's text:

> But there are several important ways in which this observation must be qualified. First his notebooks (and his notes) show that he had or almost had the same idea a number of times before ... so the historic moment was in a sense a re-cognition of what he already knew or almost knew. Second, the moment is historic more in hind-sight than it was at the time. After having the idea, Darwin reverted to other preoccupation. It took more time. ...[3]

No one to my knowledge has analyzed episodes in Darwin's intellectual development more concisely, perspicaciously, and insightfully than has David Kohn. I would urge him to consider filling the gaps and writing a book which would address all of Darwin's intellectual development.

Notes

1   David Kohn, "Theories to Work By: Rejected Theories, Reproduction, and Darwin's Path to Natural Selection," *Studies in History of Biology* 4 (1980): 67–170.
2   S. S. Schweber, "Darwin and the Political Economists: Divergence of Character," *J. History of Biology* 13 (1980): 195–289.
3   Howard E. Gruber, "Cognitive Psychology, Scientific Creativity and the Case Study Method," in: *On Scientific Discovery*, ed. M. D. Grmck, R. S. Cohen and G. Cimino (Dordrecht/Holland: Reidel 1980), pp. 295–322.

Molecular versus Biological Evolution and Programming

HENRI ATLAN

In 1971 M. Eigen published a long paper,[1] launching a series of publications which has provided a sound chemical kinetic basis for a neo-Darwinian theory of prebiotic chemical evolution. At least two great contributions of Eigen's theory to the understanding of biological evolution must be mentioned at the outset.

First, it gave a clear meaning to and a noncircular definition of selective pressure and natural selection for the first time, by providing a definition of the selective value of information-carrier molecules based on chemical kinetics. Otherwise we are left with the known circularities of selection by "survival of the fittest," where fitness is defined as ability to survive.

Second, in opposition to the thesis of Jacques Monod in his book *Chance and Necessity*, the appearance of life is not viewed as a highly improbable phenomenon which has occurred once, but rather as a necessary phenomenon of complexification and self-organization of matter, resulting from the work of physicochemical laws on the kinetics of self-reproducing polymers. However, although the appearance of complex molecular structures leading to living organisms is necessary, their particular specific structures are due partly to randomness which is introduced by an unavoidable rate of random errors. As such, randomness plays the double role of destroying specificity on the one hand, and of increasing complexity

137

E. Ullmann-Margalit (ed.), The Kaleidoscope of Science, 137–145.
© *1986 by D. Reidel Publishing Company.*

on the other. This result is basically the same as that of a principle of self-organization by complexity from noise (rather than order from noise), on which I myself have worked [2-4] in order to account for the logic of complexification observed in ontogenesis and phylogenesis.

What I would like to do here is to compare Eigen's theory of molecular prebiotic evolution with what is known about biological evolution at the molecular level, i.e. within extant biological species. We shall see that, whereas Eigen's theory is strictly neo-Darwinian, an analysis of the data on molecular biological evolution may lead to new conceptions about evolution, some of them non-Darwinian. The by-now classical neo-Darwinism thus requires rather drastic modification.

M. Kimura[5] has initiated such an analysis by comparing the molecular evolution of given proteins, found in a wide range of species, with the data of paleontologists and zoologists on the evolution of the species themselves. For example, hemoglobin is found in almost all vertebrates, and its a-chain is made up of 141 amino acids. Of these amino acids in man only 16 differ from those in mice, 18 from those in horses, and 68 from those in carp.

The earlier two different species diverged in evolution, the greater the number of differences between them. On the basis of this kind of data, it is possible to calculate the rate of change of amino acids in a given protein throughout the evolution of the species. It is striking that this rate is constant for a given protein throughout the evolution of all tested species, irrespective of the rate of evolution of the individual species as estimated from changes in phenotypic characters, such as the size of the brain, the shape of the feet or the wings, pigmentation, etc.

In order to explain this finding one must take into account the fact that the phenotypic characters which evolve and have selective values are different from the genes that are transmitted, and that evolve in a different way, apparently independent of selective advantages. In other words, it is as if there were two different kinds of evolution superimposed on one another: one which comes about by natural selection, but involves only the phenotypic characters; the other which involves the genes but occurs completely at random, unaffected by natural selection.

The problem is that it is only the genes that are transmitted from one generation to the next, not the phenotypic characters. This is why the study of the mechanisms of evolution must take both into account. And the discrepancy between their rates of evolution renders the simple scheme of neo-Darwinism untenable. This discrepancy is due to the fact that the relationship between genotype and phenotype is not a simple bijection, one gene—one character. The way genes determine characters is not straightforward; it is not a simple correspondence as in the relationship between the sequences of DNA nucleotides and protein amino acids.

Several genes and the environment may jointly produce a given character; and a single gene may also be involved in the determination of several characters. Neo-Darwinism was based on the idea of a one-to-one relation between a gene and a character. Its underlying idea was that mutations occur randomly and produce new genes. The new characters which derive from them have some selective value in a given environment as compared with other characters produced by different competing genes. Depending on this selective value, the mutant gene will either be selected or not, after a given number of generations.

Unfortunately, this relationship is not so simple because the determination is not one-to-one but more complex: several genes determine several characters. Moreover, the respective roles played by the environment and the genes in what is called epigenesis, i.e. the development leading to the constitution of the phenotype (the ensemble of characters), are not yet clear. It is not, as one might think, similar to a straightforward determination by a computer-program, based on the metaphor of the genetic program; we shall return to this point later.

This results in a discrepancy between the kind of evolution which is observed at the level of the phenotypes, and that observed at the level of the genes. Changes in the amino acid sequence of a protein depend directly on changes in the nucleotide sequence of a gene. Thus the evolution of the molecular structures represents the evolution of the genes themselves, i.e. of what is transmitted. And the measurements of rates of evolution give the result I have mentioned, according to which the mutations at the molecular level are neutral, i.e. they are produced and selected randomly, none having a higher survival value than others. More precisely, the only selection taking place is the elimination of the nonviable mutants. But no significant differences between selective values can be detected among all the others. And this explains the fact that the rate of change of a given polypeptide, measured as a percentage of amino acid substitutions per million years, is always the same for this polypeptide, from the most ancient species up to the most recent. In contrast, the rate of evolution of phenotypic characters differs considerably in different species: the size of the human skull has more than doubled in less than 2 million years, whereas old species like coelacanths or lingulae seem to have retained the same shape for 200 million years or more.

In other words, it is true that *phenotypic characters* may offer a survival advantage to the individuals who possess them, but the *viable genes* are transmitted as if they were neutral, i.e. with no survival advantage. This was shown by Kimura in the following very simple calculation: when a new viable gene appears in a population of N individuals with 2 chromosomes each, it is present as one amongst $2N$ different genes. Assuming that it has the same survival value as every other gene, the probability of its being fixed and selected in a given species is $1/2N$. For a given polypeptide, let us call μ the probability of mutation of one amino acid at a given site per year. If n is the number of amino acids in the chain, and α the fraction of viable mutants, the probability of the appearance of new amino acids in the polypeptide is $\alpha n \mu$. And the total number of new amino acids in this particular peptide within the whole population is $2N\alpha n\mu$.

Therefore the probability of a new amino acid being fixed is

$$\frac{1}{2N} \times 2N\alpha n\mu = \alpha n\mu$$

$= $ Constant, independent of the number of individuals in the species, depending only on the rate of viable (but neutral) mutations in a given polypeptide chain.

In other words, the evolution of the polypeptide chain is due only to random mutations and random fixations among equally viable mutants, no account being taken of any higher or lower selective value which would be determined by its effects on the phenotype.

All this has led people involved in what is called population molecular genetics[5,6] to reconsider the neo-Darwinian theory and to propose non-Darwinian theories of evolution, based on the occurrence of neutral mutations with natural selection playing a much less dominant role.

Darwin stated his theory by saying that "individuals who have any advantage over others have a better chance to survive and to transmit their own type." What is definitely wrong with this view is that the individuals do not transmit their type. Even the very fact of sexual reproduction prevents them from doing so, since they transmit only half of their genotype, and their offspring will be new original individuals. According to neo-Darwinian theories, following Mendel's laws and molecular genetics, this difficulty was overcome by taking into account the probability of parental gene associations in the offspring; but this was possible only on the assumption that selective advantages of phenotypes would be the expression of a selective advantage of the transmitted genes. We have seen that this is not the case. And we are left with non-Darwinian theories limited to a neutral evolution of molecular structures with no obvious selective pressure acting on them.

If one is not satisfied with this and wishes to preserve a general theory of evolution that takes natural selection into account, one has to complicate seriously the neo-Darwinian theories by considering that natural selection does not act on isolated genes, or even on a couple of genes, but on whole individuals. And the individuals are the results of nonlinear interactions between a collection of genes consisting of tens of thousands units on the one hand and their environment on the other. The mathematics of such models becomes very complicated but can be solved by computer simulations which provide quite unexpected results. Because of the nonadditivity of the effects of different genes and the nonlinearity of the equations. several different solutions are possible for the same kinetics, and the evolution of a given population depends critically on initial conditions. For example, one gene can be eliminated by natural selection in a given population and be fixed and become dominant in a neighboring population of the same species in an identical environment. Moreover, as was shown by Lewontin,[7] in contradiction to the basic law of neo-Darwinism, the effect of natural selection on a population may very well be an overall decrease in the average selective value of that same population. For all these reasons it is difficult to transpose Eigen's model of molecular evolution, which is strictly neo-Darwinian, to biological evolution; this is somehow ironical, since neo-Darwinism was first propounded for biological evolution! However, it is very valuable as a model of self-organization at the level of the individual, i.e. for ontogenesis.

In fact, as we have seen, the main difficulties are encountered in comprehending the mechanism of ontogenesis, i.e. how a genotype produces a phenotype, and it will not be possible to understand the mechanisms of evolution until we have mastered this. The commonly accepted idea is that genetic information is transmitted from generation to generation in the genome, and that it functions very much like a computer program in determining the development and the phenotype of the individuals in a given environment. That is why it is called a "genetic program." However, although the metaphor of the genetic program has proved useful in the history of modern biology, one must recognize that it should not be taken too seriously. And it is in this field of development and ontogenesis that Eigen's theory and his computer games provide valuable alternatives to plain deterministic computer programming, together with other theories of self-organization in which randomness plays an important role.

There are several reasons why the metaphor of the genetic program should not be taken too seriously. A computer program is fed into a machine from the outside and that is why the DNA molecules coming from the parents are considered to be the vectors of the program. But there is nothing in the DNA structure which makes it like a computer program, since it contains nothing approximating a programming language. One should not make the mistake of considering the genetic code as a computer language: it is at most a lexicon, with no syntax and hardly any semantics. The set of instructions which makes up a program should be sought not in the DNA structure but in the mechanisms of gene regulation by which the expression of the genes is regulated. And these mechanisms are to be found in the metabolism of the whole cell. That is why, from the very outset, the genetic program metaphor has been circumscribed by expressions such as "a program which needs the products of its reading and execution to be read and executed" or "a self-programming program."

When we consider something like a self-programming program, we encounter the main problem confronting information theory and linguistics, namely that of the mechanisms by which new meaningful information can be created. When the information content of macromolecules is computed according to Shannon's information theory, we must bear in mind that this is a measure of information based only on probabilities of occurrence of amino acids or nucleotides in polymer chains, like probabilities of occurrence of letters in sentences, without taking into account the meaning of the information transmitted. But this does not imply that such meaning does not exist. As a matter of fact, the meaning of genetic information is nothing but the phenotype of the mature organism. For example, in a cell, probabilistic information theory can be applied by noting the accuracy of the transmission of information in the channel from the DNA nucleotide sequences to the protein amino acid sequences. And up to this point there is no need to take into account the meaning of information. But such meaning does exist, and is the effect of the transmission of information on the receiver, the receiver being here the cell metabolism itself: the cell will function better or worse, depending upon the occurrence of errors in transmission and the location of these errors in the protein chain. In other words, the genetic information transmitted in that channel has a meaning for the cell, or even for the whole organism, and the effects of errors will be either to render it meaningless, or to change its meaning into a new one, as occurs in adaptation.

Thus, along the lines followed by linguists and cognitive scientists, it can be shown[4] that also in nonhuman communications, such as in these molecular communication channels, the meaning of the message is greatly dependent on the receiver and its capability to derive new meanings from random changes in the message.

Eigen has used his theory to simulate the appearance of meaningful sentences by repeating sequences of letters with errors introduced into each repetition (see Figure 1 taken from one of his works).[8]

Since I myself have devoted some time to this problem of genetic programming and creation of meaning of information, I want to conclude with a few words on a related question that I have tackled, together with Maurice Milgram.[9] The question is as follows: If we consider a network of connected cells like the central nervous system or the immune system that has evolved in the individual from the division of one single initial cell, which kind of program should be contained in this initial cell to specify the structure of a final network made up of billions of cells with hundreds of billions of connections?

If we assume that the cells behave like deterministic programmed automata, the number of instructions will be exceedingly large, and much larger than the amount of information encoded in the physicochemical states

Initial Sequence: YEKNMADSTVEXVFOMLEFT

Quality factor per symbol: q: 0.995.
Selective advantage per symbol: 10.

1. GENERATION :	ZEKNQADSTVEXFFOQLEFT
5. GENERATION :	PELNPAUSDTUTJFGI?EFV
10. GENERATION :	PEPN AUKDCETJFGJLENN
17. GENERATION :	LERN AUS DEN FEHLERN

L E R N   A U S   D E N   F E H L E R N
10001010101011001100101001010011011100001010

("Lern aus den Fehlern" means: Learn from mistakes)

Figure 1. Eigen's game of life. Starting from an initial sequence with no meaning, a meaningful sentence (target-sentence) is generated by "natural selection" (in the computer simulation, each letter, including spaces and punctuation signs, is coded by five binary digits; the information content of the sentence is 100 bits). The game proceeds as follows: the computer copies each sequence in a box with the indicated accuracy (q = 0.995 means that the error rate is 0.005, so that, on average, one binary digit out of 200 is copied with a mistake). Simultaneously, sequences disappear after a given average period of time, so that the number of sequences per box remains constant. The computer compares each sequence with a target-sentence and computes the number of symbols correctly produced. The velocity of reproduction is proportional to that number. The selective advantage indicates by how much the velocity is increased when an additional fit is obtained. In every generation there exists a dominant sequence together with mutants containing one or two errors. The choice of a target-sentence is only of significance in order to appreciate the result. This game would proceed the same way, qualitatively, if the target-sentence had not been provided in advance, but would have emerged from its meaning.

of the initial cell. The situation is completely transformed and the problem becomes tractable, if we assume the automata to be probabilistic instead of deterministic. In a deterministic automaton the transitions from one state to another, when a given input is received, are governed by rigid laws which make up its program. As against this, a probabilistic automaton will pass from one state to another only with a certain probability and may go to a different one with some other probability.

As a very simple example, let us consider, as the final structure of the network, a chain of length K, i.e. a number K of cells or automata connected sequentially in a chain. If one has to produce that structure, starting from one automaton which divides and connects itself with the following automaton, and so on, and if one assumes these automata to be deterministic, the initial one must contain at least K states, because it must work like a counter, counting up to K. If K is very large, so must be the number of states.

If, on the contrary, one assumes the automaton to be probabilistic, one can show that a chain of length K can be produced with K as large as one wishes and only five states in the automaton. Of course, the length K is not produced accurately; it is produced as an average within some range of dispersion. In other words, although the use of probabilistic automata does not lead to an unstructured chaos, one has to compromise and pay a price in accuracy for the reduction in the number of necessary states. In the same way one can obtain wheels instead of chains, and complicated trees by combining chain-producing and wheel-producing probabilistic automata into an eight-state automaton.

It is interesting to study the kind of compromise which can be achieved between the reduction in the number of states of the initial automaton and the loss of specificity in the resulting network, when we use a probabilistic automaton instead of a deterministic one. Of course, maximum accuracy with zero dispersion would be produced by a deterministic automaton. However, the difference in the number of necessary states between the deterministic and the probabilistic automaton in order to achieve a given accuracy is dependent on a third parameter, which is the complexity or lack of redundancy of the resulting network. Thus, the specificity appears as a superimposition of two features: the accuracy of the process, and the complexity of the resulting network. Hence one can understand why it is easier to accurately produce *redundant* structures with probabilistic automata. The production of diversity may be left to a second step, whereby the initial redundancy of the network would be destroyed also by random perturbations, as seems to be the case in the development of some parts of the CNS and of the immune system.

In any event, a clear advantage of this kind of game is to allow for a change of perspective concerning the so-called genetic program. One of the features which drastically increase the specificity of the resulting network (for any probabilistic automaton) is the effect of the inputs on the transition probabilities of the automaton. When applying this result, we can see the development of an organism not so much as the result of deterministic programming, but rather of a chemical automaton ruled by probabilities of interactions. The specificity of the end product is not very high to start with, since it is more or less the same in all species, i.e. it is the biochemical network of metabolic reactions. However, the importance of the inputs for increasing drastically the specificity becomes evident if we consider DNA not as a program but rather as the inputs to probabilistic chemical automata. It seems to me that this change in perspective, from viewing DNA as a program to regarding it as inputs to a relatively nonspecific probabilistic chemical automaton, is a step forward in the understanding of the logics of biological development.

This aproach differs from Eigen's, but is based on the same philosophy of showing how the positive role of randomness must be taken into account to explain evolution in time of complex natural organizations not only at the level of the evolution of the species, but also at that of the development of the individual, viewed along the lines of Piaget, as a self-organizing or nondirected learning system.

Notes

1   M. Eigen, "Self-Organization of Matter and the Evolution of Biological Macromolecules," *Die Naturwissenschaften* 58 (1971): 465–523.
2   H. Atlan, "On a Formal Definition of Organization," *Journal of Theoretical Biology* 45 (1974): 295–304.
3   H. Atlan, "Sources of Information in Biological Systems," in: *Information and Systems*, ed. B. Dubuisson (New York: Pergamon Press, 1978), pp. 177–184.
4   H. Atlan, "Hierarchical Self-Organization in Living Systems: Noise and Meaning," in: *Autopoiesis: A Theory of Living Organization*, ed. M. Zeleny (New York: North Holland, 1981), pp. 185–208.
5   M. Kimura and T. Ohta, *Theoretical Aspects of Population Genetics* (Princeton: Princeton University Press, 1971).
6   A. Jacquard, *Genetics of Human Populations* (San Francisco: Freeman, 1978).
7   R.C. Lewontin, *The Genetic Basis of Evolutionary Changes* (New York: Columbia University Press, 1974).
8   M. Eigen and P. Schuster, *The Hypercycle — A Principle of Natural Self-Organization* (Heidelberg: Springer Verlag, 1979); M. Eigen, "L'émergence des langages de la vie," *Prospective et Santé*, no. 18 (1981): 7–19.
9   M. Milgram and H. Atlan, "Probabilistic Automata as a Model for Epigenesis of Cellular Networks," *Journal of Theoretical Biology* 103 (1983): 523–547.

Gamow's Theory of Alpha-Decay

ROGER H. STUEWER

Introduction

George Gamow burst upon the European community of physicists like a meteor from outer space. The origin of his trajectory was distant Leningrad; his point of impact was Göttingen; the time was mid-June 1928. The impression Gamow made has been recorded by Léon Rosenfeld. "I shall never forget," Rosenfeld recalled, "the first time he appeared in Göttingen — how could anyone who has ever met Gamow forget his first meeting with him — a Slav giant, fair haired and speaking a very picturesque German; in fact he was picturesque in everything, even in his physics."[1] Gamow had learned German from a private tutor as a youth in Odessa with the result, he later recalled, that "I'm terribly poor in *der, die, das,* and my grammar is horrible, but pronunciation good."[2]

Gamow not only projected himself into an unfamiliar city in June 1928; he projected himself into an unfamiliar field of physics, nuclear physics, with consequences as spectacular as his personality. For he discovered, immediately, that he could explain the emission of α-particles by radioactive nuclei. And that theory, as Hans Bethe stated in 1937, "was the first successful application of quantum theory to nuclear phenomena."[3] Bethe's simple yet eloquent characterization encapsulates the deep reorientation in nuclear physics that I shall explore in this paper.

147

E. Ullmann-Margalit (ed.), The Kaleidoscope of Science, 147–186.
© *1986 by D. Reidel Publishing Company.*

For, although new to Gamow, nuclear physics was far from a new field of research by 1928. Radioactivity had been discovered in 1896, the nuclear atom proposed in 1911, artificial disintegration discovered in 1919. If nuclear physics was still in its infancy by 1928, it was an infancy filled with rich and formative experiences. My first task, therefore, will be to sketch significant developments in nuclear physics prior to 1928, to establish the problem situation at that time. I will then examine the nature of the new theory of α-decay. Finally, I will discuss its reception and influence on nuclear physics.

Nuclear Physics Prior to 1928

The first convincing experiment establishing the nature of the α-particle was carried out in 1908 in Manchester by Ernest Rutherford and T. Royds, who proved that it consisted of a doubly-ionized helium atom.[4] That it behaved like a point-charge in scattering experiments was an essential assumption leading Rutherford to his conception of the nuclear atom in 1911.[5] It then followed that the α-particle was the nucleus of the helium atom and, furthermore, that it was a constituent part of radioactive nuclei, since their decay could not be influenced by ordinary physical and chemical means.

The case of the β-particle was not nearly so straightforward. While it was soon known that β-particles were identical to ordinary electrons, Otto v. Baeyer, Otto Hahn, and Lise Meitner working in Berlin proved, in 1911–1912, that they were emitted by thorium active deposit and other radioactive elements in groups of definite velocities.[6] This and other evidence led Rutherford to conclude, in August 1912, that the emitted β-particle, as well as γ-rays, originated in "the instability of the electronic distribution."[7] Rutherford changed his mind on this point only after Antonius van den Broek, a Dutch lawyer and amateur scientist, argued, in November 1913, that if the nuclear charge Z of a particular element is identical to its atomic number, and if the nucleus consists mostly of doubly charged α-particles, "then the nucleus too must contain electrons to compensate this extra charge."[8] Rutherford was prepared to accept van den Broek's hypothesis, because by this time he and H. Robinson had proved that β-rays and γ-rays produce far too much heat energy to be attributed to the electronic distribution.[9] In addition, he soon argued that β-particles had to join α-particles as nuclear constituents to account for the radioactive displacement laws, for the existence of isotopes, and for other phenomena.[10] The entire situation, however, grew much more complicated once again in 1914 when James Chadwick, then working in Berlin, proved that the β-ray spectrum of

Radium B + C consists of a *continuous* spectrum superimposed upon its previously observed *line* spectrum.[11]

The story of how the electron, once it had been assumed to exist in the nucleus, became universally incorporated into numerous models of the nucleus proposed during the 1910s and 1920s; of how this nuclear electron hypothesis existed side-by-side after 1926 with fundamental theoretical objections to it; of how it survived Chadwick's discovery of the neutron in 1932; and of how it finally fell in 1934, is a fascinating one which I have treated in detail elsewhere.[12] In this paper I must restrict myself to those aspects of it that bear directly on the general question of the structure of the nucleus prior to 1928.

By far the most significant new information that was obtained on nuclear structure followed from Rutherford's discovery of artificial disintegration, which he reported in April 1919,[13] just before he left Manchester for Cambridge to become J.J. Thomson's successor as Cavendish Professor. Rutherford found, after a long series of experiments, that RaC α-particles incident on nitrogen ejected long-range particles (maximum range about 28 cm in air) which were "probably atoms of hydrogen." "If this be the case," Rutherford wrote, "we must conclude... that the hydrogen atom which is liberated formed a constituent part of the nitrogen nucleus" (p. 589). Thus, Rutherford concluded that hydrogen atoms — or better, hydrogen nuclei or protons — had to join electrons and α-particles as independent nuclear particles.

The enormous importance of Rutherford's discovery soon led to an invitation to deliver a Bakerian Lecture — Rutherford's second — before the Royal Society on June 3, 1920.[14] By that time Rutherford's thoughts on nuclear structure had progressed still further. Recognizing that the α-particle could be considered to be composed of 4 protons plus 2 electrons, and thinking that he also had uncovered evidence for another nuclear constituent, an X^{++} particle, which could be considered to be composed of 3 protons plus 1 electron, Rutherford was led to speculate, ultimately, that 1 proton could also bind 1 electron, that is, could form a neutron (p. 34). This conviction prompted him to encourage a number of experiments in subsequent years designed to find evidence for the existence of neutrons. It also was a crucial influence on Chadwick's thought, not only because it facilitated his recognition and actual discovery of the neutron in 1932, but also because it supported his belief, based upon his mass-energy calculations, that the neutron was not a new elementary nuclear particle but rather an electron-proton compound with a binding energy of 1–2 MeV — a basic reason for the survival of the nuclear electron hypothesis after 1932.[15]

Instead of pursuing this momentous chain of events further, I wish to emphasize another aspect of Rutherford's thought on nuclear structure. Thus, already in his initial paper of 1919, he suggested a detailed model for the structure of the nitrogen nucleus to account for its disintegration under α-particle bombardment. Noting that the mass of nitrogen-14 is of the form $4n + 2$, where n is an integer, Rutherford suggested that its constituent hydrogen nuclei or protons were "outriders of the main system of mass 12" which, when expelled by the incident α-particle from various points in their orbit, would give rise to their observed velocity distribution.[16] Rutherford soon drew a picture of this process, as shown in Figure 1.[17] Clearly, the residual

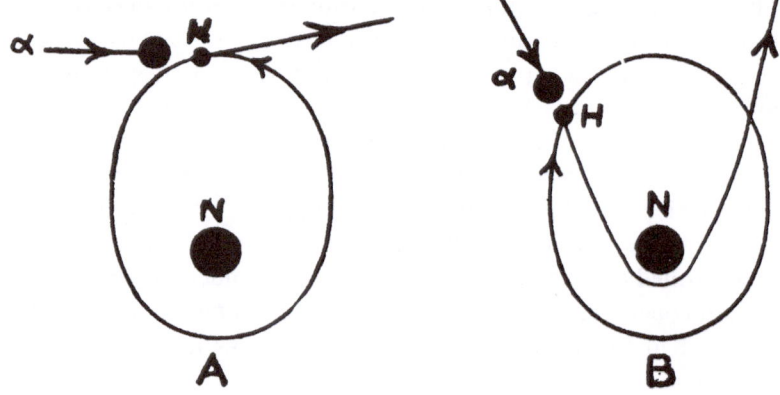

Figure 1. Rutherford's satellite model of the nucleus. An α-particle is shown ejecting a proton satellite (H nucleus) from two points in its orbit about the nitrogen core.

nucleus then is of atomic number 6 and mass 13, in other words, an isotope of carbon. That the incident α-particle is actually *captured* by the nitrogen nucleus, leaving as a residual nucleus an isotope of *oxygen*, only became clear after P.M.S. Blackett obtained cloud-chamber photographs of the disintegration process at the end of 1924.[18]

Meanwhile, Rutherford, assisted by Chadwick, bombarded element after element with α-particles, finding, by mid-1922, that all elements of odd atomic number between 5 and 13, whose isotopic masses were of the form $4n + 2$ ($^{10}_5B$, $^{14}_7N$) or $4n + 3$ ($^{11}_5B$, $^{19}_9Fl$, $^{23}_{11}Na$, $^{27}_{13}Al$), could be disintegrated, while the intervening elements of even atomic number, whose isotopic masses were of the form $4n$ ($^{12}_6C$, $^{16}_8O$, $^{20}_{10}Fe$, $^{24}_{12}Mg$), could not.[19] These data clearly supported Rutherford's idea that the incident α-particles were ejecting proton "outriders." It also encouraged him to develop a detailed, quantitative "satellite" model of the nucleus, which he pursued and refined relentlessly in succeeding years. Rutherford definitely revealed himself to be a "crypto-theoretican," as Maurice Goldhaber recently characterized him.[20]

I must leave a full account of Rutherford's satellite model of the nucleus to another paper; I can only emphasize certain of its significant features here. First, it is noteworthy that neither Rutherford's theory, nor his and Chadwick's experiments, went unchallenged. In 1922 Hans Pettersson and Gerhard Kirsch, working in Stefan Meyer's Institut für Radiumforschung in Vienna, opened up what K. K. Darrow termed the "most famous controversy" of the day with Rutherford and Chadwick.[21] Searching for relatively short-range secondary protons, Pettersson and Kirsch eventually claimed that many more elements could be disintegrated under α-particle bombardment than those reported by Rutherford and Chadwick, and with higher proton yields. They claimed, in particular, that carbon and oxygen, both of mass type $4n$, could be disintegrated. In addition, Pettersson (the true leader of the pair) advanced an "explosion hypothesis" to account for the disintegration process and as a direct challenge to Rutherford's satellite interpretation.

The battle lines therefore were clearly drawn between Vienna and Cambridge, and numerous shots were fired, both publicly in the literature and privately in correspondence, during succeeding years.[22] Eventually, at the end of 1927, Chadwick was dispatched to Vienna to witness the opposition's observations personally. He found, to his astonishment, that Pettersson had turned the extremely tedious and tricky matter of actually observing the scintillations produced by the secondary protons over to several young female assistants — who had foreknowledge of what they were expected to observe.[23] It is pleasant to report that, although Pettersson became furious in Vienna when Chadwick exposed this psychological effect, he and Chadwick subsequently became close personal friends.[24] Undoubtedly, however, the most positive consequence of the entire controversy was that it prompted Rutherford to support the development of electrical counting techniques at the Cavendish Laboratory to replace the venerable scintillation method — and the human observer.[25]

Throughout the controversy, Rutherford, of course, never doubted whose experiments were correct. Nor did he falter in his interpretation of them. Quite the contrary: Rutherford's confidence in his satellite model of the nucleus increased as he and Chadwick uncovered new experimental evidence that could be interpreted by it. Thus in July 1925 Rutherford and Chadwick reported[26] a striking decrease in ordinary Coulomb scattering when α-particles of increasing energy were incident upon aluminum and magnesium foils — a decrease they explained by assuming that the central positive cores of these nuclei were "surrounded by a satellite distribution of positive and negative charges" (p. 162), so that α-particles possessing sufficient energy to penetrate into the intermediate negative region would be scattered less intensely than ones of somewhat lower or higher energy.

Even more significantly, however, Rutherford saw in his satellite model the resolution of a striking and highly puzzling paradox that he and Chadwick discovered at this same time in the behavior of uranium (pp. 156–158). This paradox was simple and compelling: Rutherford and Chadwick found experimentally that α-particles of relatively high energy could penetrate the uranium nucleus to a distance of about 3.2×10^{-12} cm from its center and still undergo ordinary Coulomb scattering. Yet, the uranium nucleus, being radioactive, freely emitted α-particles of much lower energy, corresponding to a point of emission of about 6×10^{-12} cm from its center. Why could low-energy α-particles get out of the uranium nucleus, while high-energy α-particles were simply scattered away without being able to enter it?

Rutherford first suggested that this provocative paradox could be resolved if the uranium core were surrounded by positive and negative satellites sufficiently close together to form "charged doublets" which, to incoming high-energy α-particles, would appear electrically neutral and hence would permit them to pass to a deep level before being scattered away (p. 163). Later, however, Rutherford modified his interpretation. Encouraged by calculations of P. Debye and W. Hardmeier, which proved that electrical polarization forces could hold neutral satellites in orbit,[27] Rutherford suggested that the uranium core was surrounded by neutral satellites consisting of ordinary α-particles which had somehow acquired two neutralizing electrons.[28] *Incident* high-energy α-particles then would be unaffected by these neutral a-satellites and hence could penetrate deeply to about 3.2×10^{-12} cm, while radioactive *a-decay* would occur when these neutral α-satellites, circling at about 6×10^{-12} cm from the center of the nucleus, somehow lost their two neutralizing electrons to the nuclear core and were propelled away by the large positive central charge.

Rutherford developed this theory in detail and quantitatively in a paper entitled "Structure of the Radioactive Atom and Origin of the α-Rays," which he published in the *Philosophical Magazine* in September 1927.[29] He envisioned the nucleus of a heavy atom such as uranium to consist of three distinct regions: (1) the central nuclear core extending to a distance of about 1 $\times 10^{-12}$ cm; (2) the region between 1×10^{-12} cm and about 1.5×10^{-12} cm which was "probably occupied by electrons and possibly also charged nuclei of small mass"; and (3) the region between 1.5×10^{-12} cm and about 6×10^{-12} cm which was "occupied by a number of neutral satellites held in equilibrium by the polarizing action of the electric field arising from the central nucleus." In the case of radioactive atoms, these were neutral α-satellites; in the case of other heavy atoms, these might be "neutral satellites of mass 2 or mass 3 and possibly even of mass 1 — neutrons..." (p. 200). Confining his detailed calculations to the former case, Rutherford set up the appropriate force equation, assumed that the neutral α-satellites moved in quantized orbits, and obtained an expression for the energy E of the emitted α-particles as a function of the atomic number Z of the residual nucleus relative to that of a reference element, RaA, and involving three adjustable constants in addition to the quantum number n.

Instead of describing in detail how Rutherford chose these constants to yield a best fit to the experimental data, and found "very fair agreement between theory and experiment" (p. 189), it is more important to point out two specific consequences of Rutherford's theory. First, it provided an explanation of the origin of γ-rays as radiations produced when the neutral α-satellites underwent quantum transitions from higher to lower energy levels. Second, it entailed a specific prediction regarding the linear variation of the logarithm of the radioactive decay constant λ and α-particle energy E — the famous Geiger–Nuttall relationship established empirically as long ago as 1912.[30] In view of the fact that the escaping α-particle, in addition to its orbital energy, acquires part of its final energy from the repulsive central field, Rutherford concluded that it is

> doubtful whether the empirical relation found by Geiger has any exact fundamental significance. On the views advanced in this paper, it is to be anticipated that the radioactive constant λ should be connected not with the final energy of escape of the α-particle, but with the quantum number n characterizing the orbit of the satellite which is liberated, and also, no doubt, with a quantity depending on the constitution of the central nucleus.[31]

Rutherford summarized his entire theory in the following words:

> We can form the following picture of a radioactive atom. One of the neutral α satellites, which circulates in a quantized orbit round the central nucleus, for some reason becomes unstable and escapes from the nucleus losing its two electrons when the electric field falls to a critical value. It escapes as a doubly charged helium nucleus with a speed depending on its quantum orbit and nuclear charge. The two electrons which are liberated from the satellite, fall in towards the nucleus, probably circulating with nearly the speed of light close to the central nucleus and inside the region occupied by the neutral satellites. Occasionally one of these electrons is hurled from the system, giving rise to a disintegration electron. The disturbance of the neutral satellite system by the liberation of an α-particle or swift electron may lead to its rearrangement,

involving the transition of one or more satellites from one quantum orbit to another, emitting in the process γ-rays of frequency determined by quantum relations (p. 201).

Rutherford's satellite model of the nucleus represented the most thoughtful analysis of nuclear structure of the period. Its consequences for understanding radioactive α-decay represented the most thoughtful analysis of this problem at the time Gamow entered the field.

Gamow's Theory of α-Decay

George Gamow was born in Odessa, Russia, on March 4, 1904, by caesarean section. His father, the son of an army colonel, taught Russian language and literature in a private secondary school for boys. His mother, the only daughter of Metropolitan Arseni, the chief priest of Odessa's cathedral, taught history and geography in a private school for girls before her death only nine years after the birth of her son.[32]

Gamow received his higher education at Novorossia University in Odessa and at the University of Leningrad, where he passed all of his examinations for his intermediate diploma in the spring of 1925. His true education in physics, however, owed far less to courses and teachers than to intense discussions with fellow students at Leningrad, particularly Dimitri Ivanenko and Lev Landau. "The Three Musketeers," as they called themselves, at times joined by others such as Vladimir Fock, avidly followed the literature and discussed it thoroughly and frequently in the Borgman Library. By the fall of 1926 Gamow and Ivanenko could coauthor a paper in the *Zeitschrift für Physik*[33] that displayed their mastery of both general relativity and Schrödinger's totally new wave mechanics.

For his thesis research, after an abortive attempt to pursue experimental optics in Professor Dimitri Rogdestvenski's institute,[34] Gamow was assigned a theoretical topic, involving an application of Ehrenfest's adiabatic hypothesis, by his new advisor, Professor Yu. A. Krutkov. Gamow had little heart for this research, however, because he sensed that the topic was already passé. Thus when Professor O. D. Khvolson, who was then retired but who had earlier recognized Gamow's talents, recommended that Gamow be permitted to spend the summer of 1928 in a new intellectual atmosphere — in Max Born's institute in Götttingen — both Gamow and his thesis advisor Krutkov enthusiastically agreed. Cheered on by numerous close friends of both sexes, Gamow, 24 years old, boarded a steamer for Swinemünde in early June, and thence took a train for Göttingen, arriving there about June 11, 1928.[35]

Gamow was one of a generation of physicists that included people like Hans Bethe and Viktor Weisskopf who were just young enough to have missed participating in the creation of quantum mechanics, capped off by the recently concluded Como and fifth Solvay Conferences in September and October 1927. Unlike most of his contemporaries, however, Gamow had little desire to compete in the elaboration of quantum mechanics. The theory was already too sophisticated mathematically for his taste, and the field itself too crowded — with highly talented people, both young and old, at that. Instead, propelled by an instinct that marked his entire career, Gamow searched for a new field to explore and enlighten. He went into the physics library in Göttingen shortly after his arrival and began to scan the recent literature. One of the journals he picked up, on his very first day in the library, was the *Philosophical Magazine*; and in that journal the article that compelled his attention was Ernest Rutherford's paper, "Structure of the Radioactive Atom and Origin of the α-Rays," published in September 1927.[36] He read Rutherford's account of the paradoxical behavior of uranium and, he later recalled, "before I closed the magazine I knew what actually happens in this case. ..."[37]

Gamow's thorough self-education program in quantum mechanics at Leningrad had given him the crucial theoretical background to view Rutherford's uranium paradox in an entirely new light. Furthermore, Gamow was unusually gifted in a particular respect. As Stanislaw Ulam observed:

> This ability to see analogies between models for physical theories Gamow possessed to an almost uncanny degree. ... [It] was wonderful to see how far he could get with the use of intuitive pictures and analogies obtained by historical comparisons or even artistic ones.[38]

It was this uncanny ability in conjunction with his solid theoretical background that enabled Gamow to see that Rutherford's uranium paradox constituted, in fact, a typical quantum mechanical tunneling phenomenon. Gamow signed his classical paper, "Zur Quantentheorie des Atomkernes,"[39] on July 29, 1928; the editor of the *Zeitschrift für Physik* received it on August 2.

Gamow opened his paper by noting that it had been "often suggested already that in the atomic nucleus non-Coulomb attractive forces play a very important role." Rutherford had concluded years ago that such forces were necessary to account for the stability of nuclei. In addition, D. Enskog recently had suggested[40] that they arise from magnetic interactions within nuclei — an idea that Enskog actually continued to develop into his own

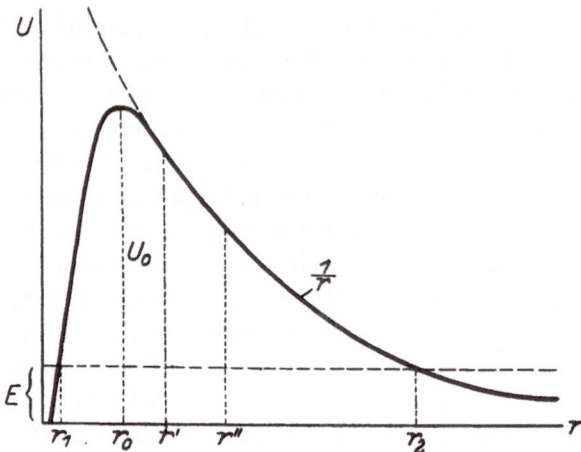

Figure 2.   Gamow's picture of the nuclear potential showing the Coulomb repulsive potential merging with the strong nuclear attractive potential. The maximum height of the potential is U_0 and the α-particle energy is E.

theory of α-decay without realizing that it was in process of being fundamentally superseded.[41] In any case, to Gamow, the existence of these attractive forces implied, as shown in Figure 2,[42] that the repulsive Coulomb potential begins to break down at some position r', that the total potential reaches a maximum U_0 at some smaller position $r_0 \sim 10^{-12}$ cm, and that in the attractive interior region $r < r_0$ the α-particle "circles like a satellite" — an unambigious reference to Rutherford's model. But that was as far as Gamow was prepared to follow Rutherford. The "principal difficulty" in the case of uranium, for example, was to explain how an α-particle of energy $E < U_0$ could traverse the region between $r_1 = 3.2 \times 10^{-12}$ cm   and $r_2 = 6.3 \times 10^{-12}$ cm, "which naturally would be impossible according to classical conceptions." Rutherford's explanation, based upon his neutral α-satellite model, "seems," Gamow bluntly concluded, "very unnatural and hardly can correspond to the facts" (pp. 204–205).

Rather, one must appeal to quantum mechanics. And with a clear appreciation for his reader's general unfamiliarity with the subject, Gamow went through, step-by-step, one tunneling problem after another. Citing recent papers by J. R. Oppenheimer and L. Nordheim, he first considered the case of a single rectangular barrier of height U_0 and width l, as shown in Figure 3,[43] with α-particles incident from the right. If the total wave function is

$$\psi = \Psi(q)\exp\left(\frac{2\pi iE}{h}\,t\right),$$

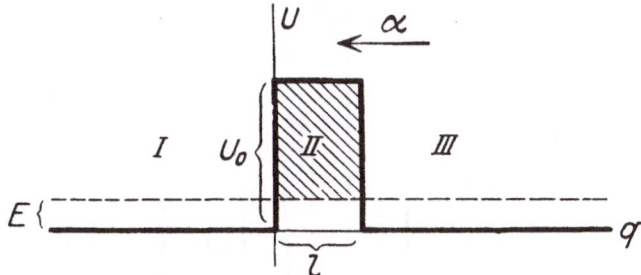

Figure 3.   Gamow's picture showing α-particles of energy E incident from the right on a single rectangular potential barrier of height U_0 and width l.

the solutions $\Psi(q)$ of the time-independent Schrödinger equation

$$\frac{\partial^2 \Psi}{\partial q^2} + \frac{8\pi^2 m}{h^2}(E - U)\Psi = 0$$

in the three regions, I, II, and III, assuming penetration from right to left, are

$$\Psi_I = A \cos(kq + \alpha)$$

$$\Psi_{II} = B_1 e^{-k'q} + B_2 e^{+k'q} \tag{1}$$

$$\Psi_{III} = C \cos(kq + \beta),$$

where $k = [(2\pi\sqrt{2m})/h]E^{1/2}, k' = [(2\pi\sqrt{2m})/h](U_0 - E)^{1/2}$, and a and β are phase factors. By applying the usual boundary conditions, i.e. by requiring the continuity of both Ψ and its first derivative $\partial\Psi/\partial q$ at both $q = 0$ and $q = l$, and introducing the assumption that the barrier is both high and broad, in other words that the product lk' is large, Gamow calculated the amplitudes B_1, B_2, and C in terms of the amplitude A. The result he emphasized was that, under such conditions, the relative transmitted amplitude A/C depends essentially on the exponential factor

$$\exp(-lk') = \exp\left(\frac{-2\pi\sqrt{2m}}{h}\sqrt{U_0 - E}\, l\right),$$

which is an extremely sensitive function of the α-particle energy E.

Gamow next turned to the case of two symmetrical potential barriers, each of width l, as shown in Figure 4.[44] Writing down the solutions Ψ_I and $\Psi_{I'}$ of the time-independent Schrödinger equation for the regions $I(q > q_0 + l)$ and $I'(q < -(q_0 + l))$, he observed that they cannot be joined smoothly at the plane of symmetry $q = 0$, because the usual *two* boundary conditions must be satisfied there, and there is only a *single* arbitrary phase factor available for

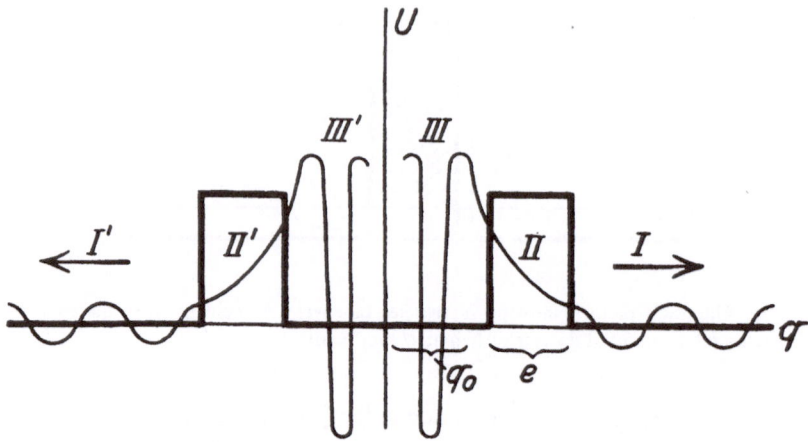

Figure 4.   Gamow's picture of two symmetrical rectangular potential barriers. The barriers are separated by a distance $2q_0$ and each is of width l (mislabeled as e).

the purpose. Physically, Gamow pointed out, this means that any solution Ψ in the interior region constructed out of these two solutions Ψ_I and $\Psi_{I'}$ would not satisfy the equation of continuity

$$\frac{\partial}{\partial t} \int_{-(q_0+l)}^{+(q_0+l)} \psi \bar{\psi} dq = 2 \left(\frac{-h}{4\pi m i} \right) (\psi \nabla \bar{\psi} - \bar{\psi} \nabla \psi). \qquad (2)$$

Thus, Gamow stated, "to overcome this difficulty we must assume that the [interior] oscillations are damped, and [hence] that [the energy] E is complex" (p. 208), i.e. that

$$E = E_0 + \frac{ih\lambda}{4\pi},$$

where E_0 is the usual α-particle energy and λ is the damping constant — or, physically, the decay constant. Now, instead of carrying through the calculation for the amplitudes in detail, Gamow observed that in physically significant cases the ratio $\lambda/(E_0/h)$ is very small (for RaC' α-particles it is $\sim 10^{-17}$), so that the variation in the wave function $\Psi(q)$ is also very small. Hence, he argued, he could take over directly the results of his single-barrier calculation by multiplying those solutions by the exponential decay factor $\exp[(-\lambda/2)t]$ and achieve his purpose — the determination of the decay constant λ — by *assuming the validity of the equation of continuity*. Thus, substituting the resulting wave functions into equation (2), Gamow calculated that

$$\lambda = \frac{4hk \sin^2 \theta}{\pi m [1 + (k'/k)^2][2(l + q_0) \kappa]} \exp\left(-\frac{4\pi \sqrt{2m}}{h} \sqrt{U_0 - El}\right), \qquad (3)$$

where κ is a number on the order of unity, and θ a phase factor. Clearly, once again, the most important result was the appearance of the exponential factor.

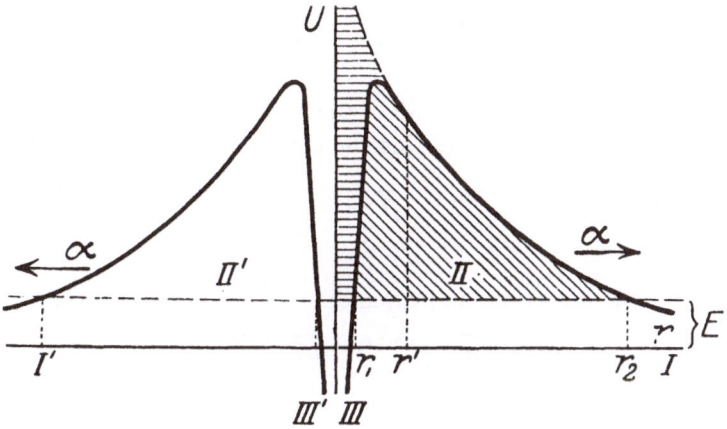

Figure 5. Gamow's picture of the "actual" nuclear potential. The energy of the α-particles is E.

Now, Gamow wrote, "we can go over to the case of the actual nucleus," as depicted in Figure 5 (pp. 209–210). This is a three-dimensional problem, so that the solution to the time-independent Schrödinger equation may be written in polar coordinates as $u(\theta, \phi)\Psi(r)$, where $\Psi(r)$ satisfies the radial Schrödinger equation

$$\frac{\partial^2 \Psi}{\partial r^2} + \frac{2}{r}\frac{\partial \Psi}{\partial r} + \frac{8\pi^2 m}{h^2}\left[E - U - \frac{h^2}{8\pi^2 m} \cdot \frac{n(n+1)}{r^2}\right]\Psi = 0. \qquad (4)$$

The azimuthal quantum number n, Gamow argued, should be set equal to zero, because the α-particle should be in its ground state before emission — indeed, he remarked, such a transition to its ground state may very well produce a nuclear γ-ray. Then, at large r, the asymptotic solution to the time-dependent radial equation is the usual spherical wave solution

$$\psi_1 = \frac{A}{r} \exp\left[i\left(\frac{2\pi E}{h}t - kr\right)\right]. \qquad (5)$$

These preliminaries aside, Gamow could proceed directly to the heart of the matter: He assumed, in complete analogy to the preceding case, a complex energy and the validity of the equation of continuity. Furthermore, he invoked an approximation, subsequently known as the WKB approximation (although Gamow neither called it that nor cited any reference to it), which enabled him to immediately write down the following expression for the decay constant λ:

$$\lambda = D \exp\left(\frac{-2\pi\sqrt{2m}}{h}\int_{r_1}^{r_2}\sqrt{U-E}\,dr\right) \tag{6}$$

where the quantity D embodies all of the unimportant factors. Splitting the area under the curve, and inserting the Coulomb potential, one has

$$\int_{r_1}^{r_2}\sqrt{U-E}\,dr \approx 2\int_0^{2Ze^2/E}\sqrt{\frac{2Ze^2}{r}-E}\,dr;$$

then, taking the logarithm and expanding to first order, one has

$$\ln\lambda = \ln D - \frac{4\pi\sqrt{2m}}{h}\left\{\int_0^{2Ze^2/E_0}\sqrt{\frac{2Ze^2}{r}-E_0}\,dr\right.$$

$$\left.+\frac{\partial}{\partial E}\int_0^{2Ze^2/E}\sqrt{\frac{2Ze^2}{r}-E}\,dr\Delta E\right\}$$

$$= \mathrm{const}_E + B_E\Delta E. \tag{7}$$

The factor B_E (the second integral in the braces) may be evaluated by inserting a change of variables, $\rho = (E/2Ze^2)r$, to obtain

$$B_E = \frac{4\pi\sqrt{2m}\,2Ze^2}{2hE^{3/2}}\int_0^1\frac{d\rho}{\sqrt{(1/\rho)-1}} = \frac{\pi^2\sqrt{2m}\,2Ze^2}{hE^{3/2}}, \tag{8}$$

which has a numerical value of 0.7×10^7 for RaA α-particles.

The final step was not as straightforward as it appears — Gamow was confronted with the above integral to evaluate. To solve this problem, Gamow later recalled,

> ... I went to see my friend N. Kotshchin, a Russian mathematician who was also spending that summer in Göttingen. He didn't believe me when I said I could not take that integral, saying that he would give a failing grade to any student who couldn't do such an elementary task. ... Later, when the paper appeared, he wrote me that he had become a laughingstock among his colleagues, who had learned what kind of highbrow mathematical help he had given me.[45]

In his final paragraph, in addition to thanking Born for allowing him to work in his institute, Gamow of course had also extended his "best thanks" to Kotshchin "for the friendly discussion over mathematical questions."[46]

In any event, Gamow's triumph was clear: In striking contrast to Rutherford, Gamow had succeeded in deriving, from fundamental principles, the linear relationship between the logarithm of the decay constant and the energy of the emitted α-particles — the venerable Geiger–Nuttall relationship. Its agreement with experiment for the uranium series is shown in Figure 6 (p. 212), where the experimental value of the

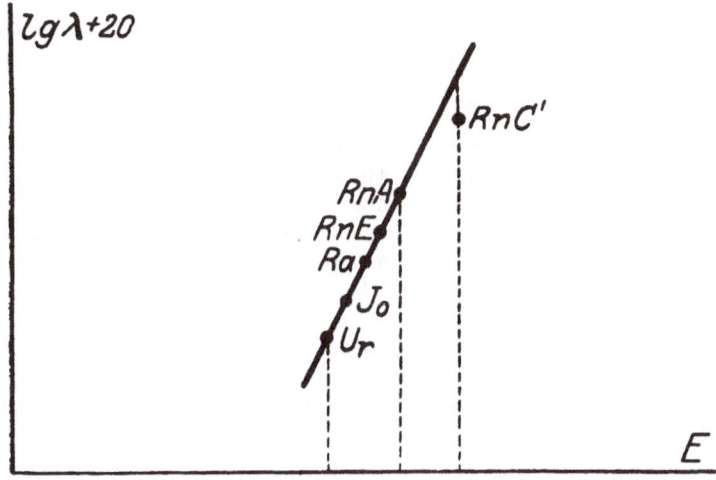

Figure 6.   Gamow's plot of the linear Geiger–Nuttall relationship between the logarithm of the decay constant (ln λ + 20 was plotted to avoid negative numbers) and the a-particle energy E. The experimental points are for elements in the uranium series.

constant B_E was found to be 1.02×10^7, of precisely the correct order of magnitude. This indicated, Gamow cautiously concluded, "that the fundamental assumptions of the theory may be correct."

The immediate reaction to Gamow's theory in Göttingen has been recorded by Léon Rosenfeld, who, as we noted earlier, was there at the time. Gamow's theory, recalled Rosenfeld,

> ... produced a remarkable sensation in Göttingen. Our feelings were rather mixed, because on the one hand we were very much impressed by the success of such a simple idea, but on the other hand we felt uneasy about it. At that time we ... had an uneasy feeling that his way of introducing those decaying states with a complex energy was contrary to the principles of quantum mechanics. I must confess that I pedantically remonstrated with Gamow that since the Hamiltonian is a hermitian operator it can only have real eigenvalues; therefore his solution with complex eigenvalues was not allowed! Gamow was quite unmoved by such scruples. He said: "Well, I have produced a solution —

you can't deny that it is a solution — which describes the phenomenon: what else do you want?" Born was still more troubled than I was, so troubled in fact that he sat down and produced a rival theory to Gamow's, using only real eigenvalues and hermitian operators. I was so pleased when I saw Born's work that I told him that now thanks to his theory I understood radioactivity.[47]

While Rosenfeld has accurately recorded Born's reaction, he also has compressed the time scale somewhat, since Born's paper[48] was not received by the editor of the *Zeitschrift für Physik* until August 1, 1929, just over a year after Gamow sent his own paper in for publication. Furthermore, we know that the first response resulting in a publication was stimulated by Fritz Houtermans, who was also in Göttingen at the time. As Gamow himself recalled:

> During my stay in Göttingen I made friends with a jolly Austrian-born physicist, Fritz Houtermans. He had recently completed his Ph.D. in experimental physics but was always quite enthusiastic about theoretical problems. When I told him about my work on the theory of alpha decay, he insisted that it must be done with higher precision and in more detail. Being a native Viennese, he could work only in a café, and I will always remember him sitting with a slide rule at a table covered with papers and a dozen or so empty coffee cups. ... We also tried to use the old electric ... computer in the university's Mathematical Institute, but it always went haywire after midnight. We ascribed this interference to the ghost of ... Gauss arriving to inspect his old place.[49]

In their resulting joint publication,[50] Gamow and Houtermans approached the decay problem as before. Introducing the change in variables $\Psi = \chi/r$, and assuming that the azimuthal quantum number $n = 0$, the radial Schrödinger equation becomes

$$\frac{d^2\chi}{dr^2} + \frac{8\pi^2 m}{h^2}[E - U(r)]\chi = 0. \tag{9}$$

The asymptotic solution to this equation, if the potential $U(r)$ vanishes at infinity, is

$$\chi_\infty = A \exp\left(\frac{-i\alpha r}{2}\right) + B \exp\left(\frac{i\alpha r}{2}\right), \tag{10}$$

where $\alpha = 4\pi \sqrt{2mE}/h = 4\pi mv/h$, $B = 0$ for outgoing waves, and A may be normalized to unity. Thus, the time-dependent radial equation

$$\psi(r, t) = \Psi(r) \exp\left(\frac{2\pi i E t}{h}\right) \tag{11}$$

has the same asymptotic spherical wave solution

$$\psi_\infty = \frac{A}{r}\exp\left[i\left(\frac{2\pi E}{h}t - \frac{\alpha}{2}r\right)\right] \tag{12}$$

as before (equation 5). Writing the equation of continuity as

$$\frac{\partial}{\partial t}\int \psi\bar{\psi}d\Omega = \frac{h}{4\pi i m}\int\left(\psi\frac{\partial\bar{\psi}}{\partial r} - \bar{\psi}\frac{\partial\psi}{\partial r}\right)d\sigma, \tag{13}$$

where $d\Omega$ and $d\sigma$ are volume and surface elements, respectively, multiplying $\psi(r,t)$ and ψ_∞ by the damping factor $\exp\left[(-\lambda/2)t\right]$, evaluating the right-hand side of equation (13), and cutting off the divergent integral on the left-hand side at some finite distance $r = R$, one obtains

$$\lambda = \frac{h\alpha^2}{4\pi m}\left(1\Big/\int_0^{\alpha R}\chi(\rho)\bar{\chi}(\rho)d\rho\right), \tag{14}$$

where a new dimensionless variable $\rho = \alpha r$ has been introduced for convenience. In terms of this new variable, and inserting the Coulomb potential $U(r) = 2Z^*e^2/r$, where $Z^* = Z - 2$ is the atomic number of the residual nucleus, the radial Schrödinger equation (9) becomes

$$\frac{d^2\chi}{d\rho^2} + \left(\frac{1}{4} - \frac{k}{\rho}\right)\chi = 0, \tag{15}$$

where $k = 4\pi Z^*e^2/hv$.

In this formulation, therefore, the problem reduced to finding a solution χ to equation (15) corresponding to (approximately) standing waves in the interior nuclear region which joins smoothly with the asymptotic solution χ_∞, and which had to be inserted into the integral in equation (14). It was Vladimir Fock — Gamow's Leningrad colleague who, like everyone it seems, was also in Göttingen that summer! — who supplied the required solution:

$$\chi \approx \sqrt{\cot u}\exp[k(2u - \sin 2u)], \tag{16}$$

where $\cos^2 u = \rho/4k$ (footnote, p. 501). Now, to facilitate the integration, Gamow and Houtermans assumed that all of the nuclear charge is concentrated in a narrow region between some definite distance $r = r_0$ and $r = R$, so that for $r < r_0$, the product $\bar{\chi}\chi = 0$. Expanding the expression for χ

about $u = u_0$, carrying out the integration $\int_{\rho_0}^{\alpha R} \chi\bar{\chi}d\rho$, substituting values for α and k, and taking the logarithm of both sides of equation (14), Gamow and Houtermans finally obtained their desired result:

$$\ln\lambda = \ln\frac{64\pi^2me^2}{h^2} + \ln Z^*v + 2\ln\tan u_0$$

$$-\frac{8\pi e^2}{h}\frac{Z^*}{v}(2u_0 - \sin 2u_0), \qquad (17)$$

where $\cos u_0 = (v/\sqrt{Z^*})(1/2e)\sqrt{mr_0}$.[51]

To compare this expression with experiment, it was necessary to fix the radius r_0. This Gamow and Houtermans did by proceeding in reverse: From the known decay constants λ and α-particle velocities v for the intermediate elements, RaEm, ThEm, and AcEm, in the uranium, thorium, and actinium series, they first calculated the corresponding values for r_0 (finding 7.35×10^{-13} cm, 7.25×10^{-13} cm, and 6.63×10^{-13} cm, respectively) and then assumed that these values held for *all* elements in each series. They recognized that this was an arbitrary procedure, but it was a common normalization technique as well. Most importantly, it enabled them to separate out each of the three radioactive series from the others, and for each to compare theory and experiment with the strictly linear Geiger–Nuttall relationship. They found — as a result of extensive numerical calculations supervised by the ghost of Gauss — "by and large right proper agreement," although they were careful to point out discrepancies as well. They noted, for example, that certain departures associated with extremely fast α-particles emitted by some elements might be accounted for if the azimuthal quantum number n were assumed to be different from zero, which would increase the apparent repulsive nuclear potential. They also noted that the effective nuclear charge Z^* would be increased if the nuclear electrons emitted in subsequent β-decays lay outside of the region in which the α-particles resided, but, they added, "since we still can make no model of the mechanism of β-emission we have not considered this hypothesis in our calculations."[52]

A Simultaneous Discovery

Gamow and Houtermans signed their joint paper in Göttingen in September 1928, just before Gamow left for Copenhagen and Houtermans for Berlin. Gamow's intention was to try to meet Niels Bohr and visit Bohr's Institute before returning to Leningrad. Much later he gave the following account of his arrival in Copenhagen:

Since my money was practically all gone, I stopped in a cheap rooming house and walked to the Institute for Theoretical Physics on Blegdamsvej. Bohr's secretary, Frøken Betty Schultz, ... said that the professor was very busy and that I might have to wait for a few days. However, when I told her that I had just enough money left to stay for one day before leaving for home, the interview was arranged for the same afternoon.

Bohr asked me what I was doing at present, and I told him about the quantum theory of radioactive decay; my paper was in the press but had not yet appeared. Bohr listened with interest and then said, "My secretary told me that you have only enough money to stay here for a day. If I arrange for you a Carlsberg Fellowship at the Royal Danish Academy of Sciences, would you stay here for one year?"

"My, yes, thank you!" I answered very enthusiastically.[53]

Gamow forgot or neglected to report that he had actually written to Bohr on July 21, 1928, proposing to visit Copenhagen at the end of August and soliciting an invitation to assist him in securing a Danish visa. Moreover, Gamow enclosed a letter of recommendation from A. Joffé to Bohr.[54] Clearly, Bohr's immediate and surprising offer of fellowship support had not occurred without some prior preparation and knowledge of Gamow's abilities.

Indeed, it is quite likely that the greatest surprise Gamow experienced, almost immediately upon arrival in Copenhagen, occurred when he picked up the September 22, 1928, issue of *Nature* and there read a letter written by Ronald W. Gurney and Edward U. Condon entitled "Wave Mechanics and Radioactive Disintegration."[55] Its opening paragraph, beautifully split by a picture of a nuclear potential, as shown in Figure 7, asserted "that disintegration is a natural consequence of the laws of quantum mechanics without any specific hypothesis." Contrary to the case of a ball confined

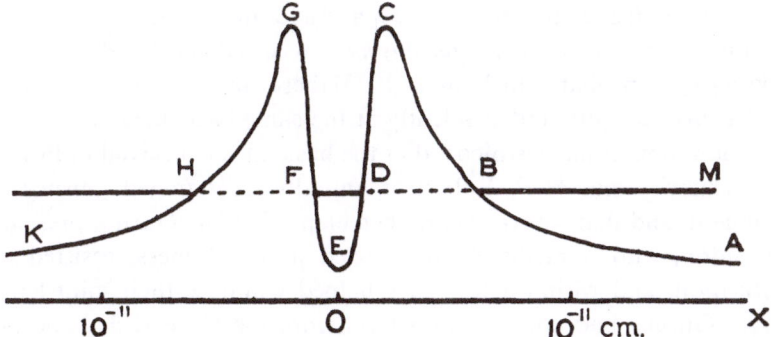

Figure 7. Gurney and Condon's picture of the nuclear potential, where the height of the line DF above the base line represents the *a*-particle energy *E*.

classically in a valley between two mountains, Gurney and Condon continued, wave mechanics allows for a "small but finite probability that the particle in the orbit DF will escape from the nucleus ...," and by varying the height of the potential barrier "through a small range we can obtain all periods of radioactive decay from a fraction of a second, through the 10^9 years of uranium, to practical stability" — the enormous variation embodied in the Geiger–Nuttall relationship. Furthermore, Rutherford's "disconcerting" uranium paradox can be resolved by this theory. Gurney and Condon concluded:

> Much has been written of the explosive violence with which the a-particle is hurled from its place in the nucleus. But from the process pictured above, one would rather say that the a-particle slips away almost unnoticed.

Thus, the quantum mechanical theory of a-decay had just joined numerous other examples of simultaneous discovery in physics. The chronology alone proves the complete independence of Gurney and Condon's work from Gamow's. Gurney and Condon signed their joint letter to *Nature* in Princeton on July 30, 1928, precisely one day after Gamow signed his article in Göttingen for the *Zeitschrift für Physik*. Gurney and Condon's letter was published, as noted above, on September 22. Gamow responded by writing a letter to *Nature*[56] on September 29, exactly one week later, but his letter was not published until November 24. Meanwhile, Gamow's *Zeitschrift für Physik* article had appeared in mid-October (the issue was closed on October 12); and Gurney and Condon had sent off a full report on their work to *The Physical Review*[57] on November 20, had described it verbally at the Schenectady meeting of the National Academy of Sciences on the same day, and had described it once again at the Minneapolis meeting of the American Physical Society on December 1.[58] Gurney and Condon's *Physical Review* article was published in February 1929; and a footnote to its title records that in "a number of the *Zeitschrift für Physik* ... received here [in Princeton] two weeks ago [i.e. probably in January 1929] there appears a paper by Gamow who has arrived quite independently at the same basic idea ...."

Condon himself has testified[59] that the basic idea conceived in Princeton was Gurney's, that H.P. Robertson initially discouraged Gurney from pursuing it, and that several weeks then elapsed before Gurney presented it to Condon, who immediately recognized its fruitfulness, assisted in its clarification, and within a few days helped compose their joint letter to *Nature*. Gurney had come to Princeton from the Cavendish Laboratory, where he had worked under Rutherford on experimental problems in radioactivity and, unlike Gamow as we have seen, had become very familiar with Rutherford and Chadwick's work between 1925 and 1927. His immediate stimulus, however, was his reading of two papers in the library of the Palmer Physical Laboratory, one by J. R. Oppenheimer,[60] the other by R. H. Fowler and L. Nordheim,[61] which independently treated the problem of field emission of electrons from metal surfaces as a quantum mechanical tunneling problem. Gurney's subsequent collaboration with Condon was a major influence in transforming him from an experimental into a theoretical physicist and productive author of textbooks.[62]

Like Gamow, Gurney and Condon were sensitive to their readers' general unfamiliarity with quantum mechanical tunneling problems. Explicitly adopting Born's probabilistic interpretation of Schrödinger's wave function, they first illustrated its use by proving that a simple harmonic oscillator in its ground state has more than a 15% chance of being outside its classically permitted amplitude. They then turned to the single-barrier problem, for which the incident, reflected, and transmitted waves obey the conservation property

$$(\Psi\bar{\Psi})_{\text{inc}} = (\Psi\bar{\Psi})_{\text{ref}} + (\Psi\bar{\Psi})_{\text{tr}},$$

and proved that for a high and broad barrier ($k'l$ large) the probability of transmission, $P_a(E) = (\Psi\bar{\Psi})_{\text{tr}} \div (\Psi\bar{\Psi})_{\text{inc}}$, is given by

$$P_a(E) = \frac{16E}{U_0}\left(1 - \frac{E}{U_0}\right)\exp\left[-\frac{4\pi}{h}\sqrt{2m(U_0 - E)}r\right], \qquad (18)$$

where the symbols have the same meaning as before. They remarked that the "nearest analogue" to this exponential variation was, "perhaps, in optics in the slight penetration of a refracted ray into a rarer medium even beyond the angle of total reflection ... ."[63] They also noted that the same factor appears for a "saw-tooth" barrier approximating the field emission problem. In both cases, they went on, the factor in the exponent can be approximated by an integral expression, to give

$$\exp\left[-\frac{4\pi}{h}\int\sqrt{2m(U - E)}\,dr\right],$$

an approximation for which they cited only G. Wentzel's 1926 paper.[64]

As a final preparatory step, Gurney and Condon considered the case of a particle in a potential valley and presented two simple arguments for estimating the factor in front of the exponential. First, if its energy is E, its velocity is $(2E/m)^{1/2}$, and hence the time it must spend in a region of length r before standing a chance of escaping is $r(m/2E)^{1/2}$ — which must be inversely proportional to the decay constant λ. Or, second, one can think of the particle as executing a motion of frequency $(1/r)(2E/m)^{1/2}$ and having a probability of escape at each impact with the barrier given by the exponential factor. Both arguments lead to the same conclusion, that the decay period $T = 1/\lambda$ is given by

$$T = \frac{1}{\lambda} \sim r \sqrt{\frac{m}{2E}} \exp\left(\frac{4\pi}{h} \int \sqrt{2m(U-E)}\,dr\right). \qquad (19)$$

It was this result that Gurney and Condon applied to the case of radioactive a-decay, and they analyzed its consequences by means of an insightful graphical technique. With Gamow they assumed a repulsive Coulomb potential and cited Enskog[65] for showing that the attractive nuclear potential might be attributed to magnetic interactions. Then, choosing an intermediate element in the uranium series, RaA, they sketched the potential barrier shown in Figure 8,[66] where the Coulomb potential was assumed to

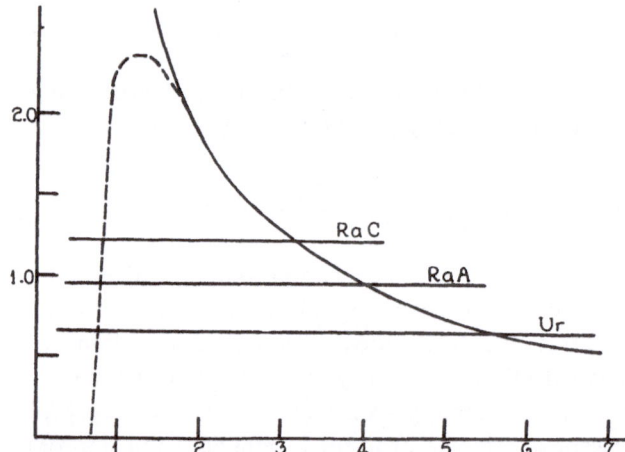

Figure 8. Gurney and Condon's picture of the nuclear potential for elements in the uranium series. The unit of ordinates is 10^{-5} ergs, of abscissas 10^{-12} cm. The line for uranium should intersect with the Coulomb potential at about 6.3×10^{-12} cm.

begin to break down at a position $r = 2 \times 10^{-12}$ cm, and where the known a-particle energy (0.955×10^{-5} erg) determined the height of the line above the $U = 0$ axis, and hence its intercept with the Coulomb barrier (at $r \approx 4 \times 10^{-12}$ cm). They further assumed, and more fundamentally, that the *same* barrier exists for *all* elements in the same radioactive series, so that they could draw similar lines for U and RaC′, knowing their α-particle energies

$(0.65 \times 10^{-5}$ erg and 1.22×10^{-5} erg, respectively — the line for U has been slightly misdrawn; its intercept should be at $r \approx 6.3 \times 10^{-12}$ cm). The area under the barrier and above these lines, however, is given by $(4\pi/h) \int \sqrt{2m(U-E)} \, dr$, and hence the above information could be displayed even better by plotting $(4\pi/h) \sqrt{2m(U-E)}$ *vs. r*, as shown in Figure 9 (p. 136). In this figure, the dashed line for RaA was determined by making the total area under the curve (53.7 units) agree with the value calculated from its known decay period (4.4 minutes); the dashed lines for RaC and U were then drawn to join smoothly with that for RaA.

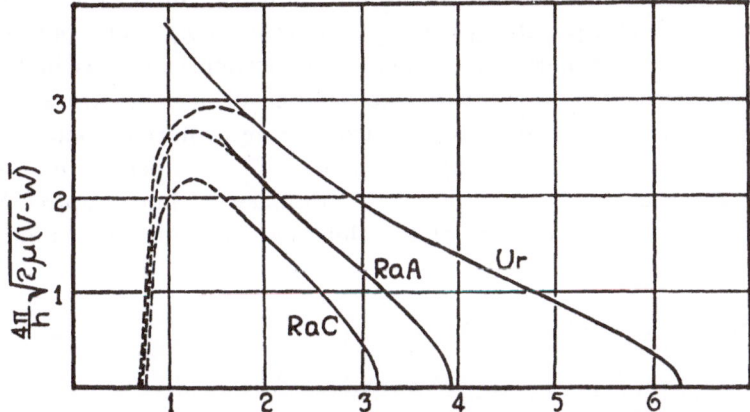

Figure 9.   Gurney and Condon's transformation of Figure 8 displaying more clearly the area under the curve for the exponential factor. (Their symbols V and W have been replaced in the text by U and E, respectively, for consistency in notation.) The unit of ordinates is 10^{13} cm^{-1}, of abscissas 10^{-12} cm.

These plots vividly illustrated Gurney and Condon's main point, for each square in Figure 9 has the dimensionless area of 10 units, corresponding to an enormous change in the decay period — a factor of e^{10}. In fact, the areas under the RaC and U curves are 34.4 and 90 units, respectively, corresponding to decay periods on the order of 10^{-6} sec and 10^{10} years — an enormous variation, and one that was in quantitative agreement with experiment, and with the Geiger–Nuttall logarithmic relationship. Furthermore, the entire theory obviated Rutherford's neutral satellite interpretation of α-decay: "[If] we abandon classical mechanics," they concluded, "the [uranium] paradox disappears, yielding us direct experimental evidence in favor of the phenomenon of quantum mechanics in which we are interested" (p. 139).

    Gurney and Condon, therefore, presented a model of α-decay identical to that of Gamow, although they differed considerably in the way they determined the decay constant. They also differed with Gamow in other respects. First, they avoided entirely the question of the origin of γ-rays and its possible connection to α-decay. Second, and in particularly sharp contrast to Gamow, they speculated that precisely the same barrier-penetration model might serve as an explanation for β-decay, a point that was criticized at the Schenectady meeting on November 20, 1928, by Irving Langmuir, who asked Condon if the theory then could explain the continuous β-ray spectrum as well as the β-ray line spectrum. Langmuir's question "upset" Condon "considerably" since, surprisingly, up to that time he had not heard of the existence of the former spectrum.[67] Finally, Gurney and Condon explicitly denied the possibility that the same basic theory could apply to the inverse problem of α-particle *penetration*, as for example displayed in P.M.S. Blackett's recent cloud-chamber photographs of the disintegration of nitrogen nuclei. In such cases, they asserted, "we are at once confronted by the fact that instead of approaching the barrier 10^{20} times per second, like a nuclear particle, our alpha-particles will only make one impact apiece. So it would seem that the capture of the alpha-particle could not be due to penetration" (p. 138).

Aftermath

Gamow's immediate response, of September 29, 1928, to Gurney and Condon's "very interesting letter," as he put it, not only contained a sketch of his own identical theory of α-decay. Gamow also argued that the "same model of the nucleus allows us to calculate an upper limit to the probability of artificial disintegration by bombardment with α-rays ...."[68] Thus, unknown to Gurney and Condon, Gamow disagreed entirely with them regarding the interpretation of Blackett's cloud-chamber photographs and other disintegration experiments. In fact, only a couple of weeks later, in October 1928, evidently after receiving some comments from R. H. Fowler and Rutherford, Gamow submitted yet another paper to the *Zeitschrift für Physik* in which he demonstrated that "we have here ... precisely the inverse of the wave-mechanical process which has ... served for the explanation of radioactive decay."[69]

To treat this problem in detail, Gamow divided the probability of disintegration into two parts, the probability W_1 that the α-particle enters the nucleus, and the probability W_2 that it then ejects a proton. The former therefore represents an upper limit on the disintegration probability, and to calculate it Gamow assumed that the nuclear potential is of the form

$$U(r) = \frac{2Ze^2}{r} - \frac{a}{r^3},$$

where the second attractive term was chosen to agree with certain experiments of E. S. Bieler,[70] and where by differentiation the constant a could be expressed in terms of the radius r_m at which the nuclear potential reached a maximum. Substituting these results into the exponential factor from his theory of α-decay and translating the three-dimensional problem into a one-dimensional approximation, Gamow concluded that the penetration probability per impact, W_1^*, was given by

$$W_1^* = \exp\left(-\frac{16\pi e^2}{h} \frac{ZJ(k)}{v_{\text{eff}}}\right) \tag{20}$$

where v_{eff} is the effective velocity of the α-particle relative to the target nucleus, $J(k) = \int (\rho^{-1} - k^2\rho^{-3} - 1)^{1/2} d\rho$, $k = mr_m v_{\text{eff}}^2/(4\sqrt{3}\,Ze^2)$, and ρ is a dimensionless dummy variable.

Equation (20) enabled Gamow to plot the probability of penetrating an aluminum nucleus as a function of α-particle range (or energy) relative to that for RaC' α-particles, as shown in Figure 10,[71] and to conclude that the results were in "fairly good" agreement with Rutherford and Chadwick's experimental values. Next, introducing Avogadro's number N and the geometrical cross-section πr_m^2 of the target nucleus, Gamow could plot the total penetration probability, $W_1 = N\pi r_m^2 W_1^*$, as a function of the atomic number of the target nucleus Z, as shown in Figure 11,[72] for two different α-particle energies, viz., for Po and RaC' α-particles. These results, too, were in good agreement with W. Bothe and H. Fränz's experiments,[73] and Rutherford's, respectively, but — Gamow emphasized — they were in "flagrant contradiction" to those "of the Viennese researchers," who had found "a fairly large yield" of disintegration protons even "for iron and other heavy elements."[74] At the same time, the fact that agreement *was* reached with the former data implied that the proton ejection probability W_2 must be nearly unity, i.e. "that (for disintegratable elements) a proton almost always will be expelled when an α-particle penetrates the nucleus."[75] The upper limit W_1, therefore, had to be close to the actual disintegration probability.

Gamow reached these conclusions just as his original paper on α-decay, and Gurney and Condon's letter, began to attract widespread attention. The near simultaneity in their appearance added emphasis to their content and precipitated reactions and counterreactions throughout the following year

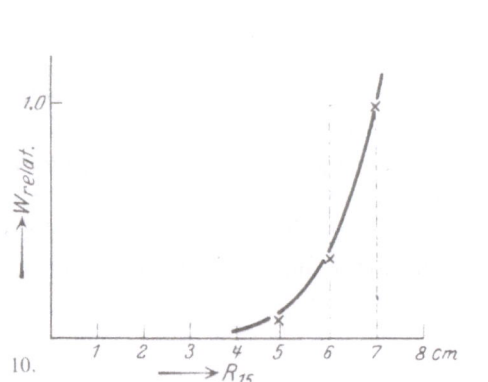
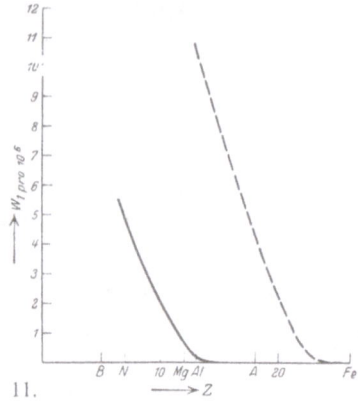

Figure 10.   Gamow's plot of probability per impact W for α-particles to penetrate an aluminum nucleus as a function of their range R (or energy), relative to that of RaC' α-particles.

Figure 11.   Gamow's plot of the total penetration probability W_1 per 10^6 impacts as a function of the atomic number Z of the target nucleus for polonium α-particles (solid line) and RaC' α-particles (dashed line).

and beyond. Max von Laue in Berlin was the first to respond, in November 1928.[76] His general assessment was widely shared.

> The quantum theory of the atomic nucleus ... establishes the first bridge from the physics of electron shells to the physics of the nucleus. Its fundamental idea ... is so illuminating, and the derivation of the Geiger–Nuttal [*sic*] relation, which until now has been completely mysterious, is so convincing, that one may consider these works as perhaps the most beautiful results of the new physics (p. 726).

Nevertheless, von Laue felt that Gamow's mathematics was "not completely in order." He questioned Gamow's solution with its complex energy eigenvalue and argued that Gamow's two arbitrary constants, the decay constant λ and the phase difference α, were insufficient to satisfy the boundary conditions at the interior plane of symmetry. He felt, moreover, that he could establish the incorrectness of Gamow's derivation of the decay constant by deriving it himself by means of a thought experiment, as follows: Imagine that one has two opposing and infinitely high rectangular potential barriers, each of thickness l and separated by a distance α, so that monochromatic waves of equal intensity would be reflected back and forth between them without diminution. Imagine further that at time $t = 0$ the heights of these barriers would be dramatically reduced, so that these waves now would be reflected back and forth in time $\tau = a/v = a(m/2E)^{1/2}$, and at each reflection would have their intensity reduced by an amount determined by the reflectivity R of the barrier — the same type of picture which, unknown to von Laue, Gurney and Condon would present in their full report.[77] Now, if J_0 is the initial intensity of the waves, after n reflections in time t, where $n\tau < t < (n+1)\tau$, their intensity would be $J = J_0 R^n = J_0 \exp(n \ln R)$, which, since the transit time τ is "unobservably small," goes over to the limit $J = J_0 \exp(-\lambda t)$, where $\lambda = (\ln R)/\tau$. In terms of the transmissivity of the barrier, $T = 1 - R$, and noting that $T \ll 1$, the decay constant becomes

$$\lambda = \frac{\ln(1-T)}{\tau} \doteq \frac{T}{\tau} = a \sqrt{\frac{2E}{m}} \, T.$$

Substituting the expression for the transmissivity of a single barrier of height U_0, under the assumption that it is both high and broad (i.e. that $k'l$ is large), von Laue found, finally, that

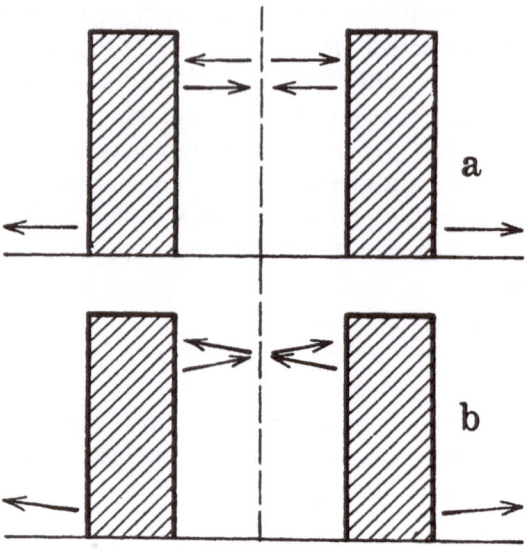

Figure 12.   Gamow's pictures intuitively illustrating why undamped symmetric solutions of the wave equation cannot satisfy both boundary conditions at the plane of symmetry (case a), while damped solutions can (case b).

the amplitudes of the incident, reflected and transmitted waves, it is evident that in the first case, which corresponds to undamped symmetric solutions of the wave equation, the arrows do not match up at the plane of symmetry, that is, the amplitudes are not equal, so that the boundary conditions cannot be satisfied there. In the second case, however, where the damping is represented by inclined arrows, the amplitudes can be made equal, that is, the boundary conditions can be satisfied at the plane of symmetry by proper choice of the constants — indeed, "the conditions to be satisfied determine the energy and associated decay constant as eigenvalues of the problem" (p. 602). And Gamow then went on to prove his point mathematically, adding in a footnote that if one were interested only in the decay constant, "one can simply use the equation of continuity, as was shown in my first article" (footnote, p. 603). In the final analysis, therefore, Gamow held his ground firmly, rejecting von Laue's calculation leading to transmitted standing waves, and not even granting that von Laue's stepwise method for deriving the decay constant fundamentally assisted in clarifying the problem.

Von Laue was only the first to react to Gamow's, and to Gurney and Condon's theory of α-decay. G.I. Pokrowski of the Physical Institute of Moscow's Technical University published a series of papers,[81] beginning in

November 1928, in which he attempted to prove theoretically and experimentally that γ-rays and X-rays can eject α-particles from a Gamow nuclear potential — results, however, which were disputed by H. Herszfinkiel and H. Dobrowolska in Warsaw.[82]

J. Kudar, who was loosely associated with Erwin Schrödinger in Berlin on a stipend from the Hungarian Ministry of Culture, was much more persistent than Pokrowski. Beginning in December 1928, Kudar wrote one paper after another at a furious rate, criticizing and elaborating Gamow's theory mathematically,[83] discussing Nernst's hypothesis,[84] and finally attempting to apply Gamow's model to β-decay.[85] Frequently, Kudar's own papers had to be corrected by others, lending credence to Schrödinger's view of him as a mathematically talented, "sympatisch," but also highly nervous and "really unlucky" man. Eventually Lise Meitner, who had a high opinion of Kudar's scientific talents, permitted him to work with her for a time in the Kaiser Wilhelm Institut für Chemie. Schrödinger, however, remained convinced that Berlin, with its highly independent theoreticians, inlcuding himself, was not the right place for Kudar and, in the spring of 1931, arranged a short visit for him to Copenhagen — where Bohr, too, liked him personally but was unable to form a deep impression of his scientific abilities.

Theodor Sexl, Professor of Theoretical Physics at the University of Vienna, dipped his oar into the waters in February 1929. Approving of von Laue's analysis, Sexl, in a series of three papers,[86] attacked Gamow, Gamow and Houtermans, and Kudar on mathematical grounds, and himself attempted to provide an exact derivation of the decay constant. These papers precipitated a detailed rebuttal by R. Atkinson and Houtermans in July 1929,[87] in which they essentially justified Gamow's original approach to the problem. Sexl returned the favor in December 1929,[88] which led Atkinson and Houtermans to respond once again — this time independently.[89] By January 1933 Sexl was still attacking Gamow's treatment in the strongest terms — he called it "thoroughly [durchaus] faulty" — and refining it mathematically.[90]

In April 1929, Chr. Møller, assisted by discussions with Gamow, O. Klein, and Bohr in Copenhagen, generalized Gamow's calculation to the relativistic case,[91] finding that both the Dirac equation and the Klein–Gordon equation led to essentially the same result, the introduction of an additional term, $-[(U - E)/c]^2$, under the radical in the exponential factor in the decay constant expression. In November 1930 Sisirendu Gupta, from the University College of Science, Calcutta, also attempted a relativistic calculation of the decay constant.[92] Meanwhile, in August 1929, Max Born presented a rigorous and extensive calculation of the decay constant based upon the "usual formulation" of quantum mechanics, which "knows only hermitian operators with real eigenvalues."[93] Five years later H.B.G. Casimir proved "that Born's treatment of Gamow's problem leads, in higher approximation, to a damped solution of the wave equation"[94] — i.e. to one consistent with Gamow's solution. To Fritz Houtermans, writing in 1930, Born's calculation simply could be "viewed as a good additional confirmation of the complex method."[95] Léon Rosenfeld, as we noted earlier,[96] found it to be more impressive than that. In general, all calculations led to the same expression for the critical exponential factor. As a result, C. F. von Weizsäcker[97] and other physicists subsequently adopted Gamow's original method simply on grounds of convenience.

The brief overview above displays, if nothing else, the great excitement generated by the appearance of Gamow's, and of Gurney and Condon's, papers. Indeed, a fair measure of their impact may be gained from the fact that Houtermans, as early as 1930, was commissioned to write an extensive review article — it ran to 99 pages — for the *Ergebnisse der exakten Naturwissenschaften* on "Neuere Arbeiten über Quantentheorie des Atomkerns."[98] By 1937 Hans Bethe could summarize the entire situation by remarking that:

> Hardly any other problem in quantum theory has been treated by so many authors in so many different ways as the radioactive decay. All the proposed methods are, of course, equivalent (insofar as they are correct) but they differ in rigor and complication.[99]

At the time, especially during 1928 and 1929, it seemed like a massive outpouring in the literature. Gamow recalled that at one point, when yet another paper had appeared, Wolfgang Pauli caustically remarked, "Es Gamowt wieder," like "Es regnet wieder."[100] It is not difficult, in at least this case, to sympathize with Pauli's causticity.

Gamow personally, apart from his reply to von Laue, held himself aloof from the mathematical criticisms and refinements of his derivation of the α-decay constant. He had conceived a physical model of the decay process; it was not his style to elaborate it further in paper after paper. He did, however, contribute materially and personally to its acceptance — by Rutherford. For it was apparent to both Gamow and Bohr that, while the new theory undercut the one Enskog began developing in 1927,[101] its principal fatality was Rutherford's, based upon his satellite model of the nucleus. Thus, not long after his arrival in Copenhagen, Gamow found himself being urged by Bohr to visit Rutherford and the Cavendish Laboratory — that fountainhead of nuclear physics.

Such a visit clearly called for a measure of diplomacy, particularly since Rutherford's commitment to his satellite model had just been strongly reinforced by Chadwick's recent visit to Vienna at the end of 1927, when Chadwick discovered that the experiments undergirding its contender, Pettersson's "explosion theory," were seriously deficient. Bohr addressed this problem by writing a letter to R.H. Fowler, the theorist in Cambridge who was closely following Gamow's work, asking Fowler to speak to Rutherford about the advantages to Gamow of a visit to the Cavendish Laboratory, and soliciting an official letter of invitation to assist Gamow in securing a visa to England. Bohr also asked D.R. Hartree and N.F. Mott, who were just returning to Cambridge after visiting Copenhagen, to pave the way personally for Gamow.[102] The results were positive: Rutherford wrote to Bohr on December 19, 1928,[103] enclosing a formal invitation to Gamow, and Gamow left Copenhagen for Cambridge on January 4, 1929.[104] To assist in the concerted effort to save him from the powerful jaws of the crocodile, as Gamow later put it,[105] he carried with him the still unpublished plots, Figures 10 and 11, demonstrating that his theory, as extended to the inverse penetration problem, agreed with Rutherford and Chadwick's experiments but disagreed entirely with Pettersson and Kirsch's. Gamow remained in Cambridge until February 12, 1929. It is amusing to note, as Mott reported to Bohr, that at first Gamow found Cambridge to be "terribly highbrow, no Dummheit machen, only terrible tea parties ... ."[106]

Rutherford received Gamow warmly. He was greatly impressed with Gamow personally. He also immediately recognized the fruitfulness of Gamow's theory of α-decay — its ability to account for the venerable Geiger–Nuttall relationship was indisputable — and as a result described it warmly in introductory remarks at a "Discussion on the Structure of Atomic Nuclei"[107] at the Royal Society on February 7, 1929, which he cordially invited Gamow to attend. Still, Rutherford had reservations. He recognized, accurately, that Gamow's theory said nothing about the detailed *structure* of the nucleus — to Rutherford it meant, as K.K. Darrow soon put it, that the "present nucleus-model consists of little more than a single curve ... ."[108] The result was that Rutherford found it difficult to reject entirely his own theory for more than a year: He presented his satellite model of the nucleus, *and* Gamow's theory of α-decay, in his, Chadwick, and C. D. Ellis's classical treatise, *Radiations from Radioactive Substances,*[109] whose preface Rutherford signed in October 1930. Nevertheless, Chadwick insisted upon also including some highly critical remarks[110] on Rutherford's model: By then it was obvious to Chadwick, and indeed to everyone else, that Gamow's (and Gurney and Condon's) theory constituted the death-knell of

Rutherford's satellite model of the nucleus.[111] The final touch came when the Cambridge Mathematical Tripos Examination for 1930 contained a question that required its takers to give an account of Gamow's theory. As Mott wrote to Bohr: "Such is fame!"[112]

Another subject that came up for discussion, particularly with R.H. Fowler, during Gamow's visit to Cambridge was the possibility of resonance — that the probability of an α-particle penetrating a nucleus might depend upon its energy being equal to some characteristic energy of the nucleus. Fowler and A.H. Wilson subsequently explored this possibility in detail mathematically in a paper[113] submitted for publication in April 1929, just as R.W. Gurney, after leaving Princeton for Tokyo, independently addressed the same question in a letter in *Nature*.[114] In contrast to Gurney, however, Fowler and Wilson concluded that their results "do not correspond to the conditions of any conceivable experiment."[115] The question was reopened one year later, in mid-1930, on theoretical grounds by Chadwick and Gamow,[116] and Atkinson, [117] and on experimental grounds by H. Pose, who, following up earlier work,[118] reported further evidence that α-particles can eject homogeneous groups of protons from aluminum foils[119] — evidence that stimulated a great deal of excitement and work in subsequent years.[120] Chadwick and Gamow had concluded that if the incident α-particle is captured by the target nucleus, the expelled protons should possess a discrete energy spectrum; if not, a continuous spectrum. This conclusion — and indeed Gamow's entire nuclear model — were adopted by W. Bothe and H. Becker a few months later as the fundamental basis for interpreting their well-known experiments in which various light elements, including especially beryllium, were bombarded with polonium α-particles.[121] They were led in this way to expect that high-energy γ-rays would be produced in the reaction — a case in which the very success of Gamow's theory may have blinded these researchers to another possibility, that another and hitherto undiscovered nuclear particle was being produced in the reaction. By the fall of 1931 Bothe was still adhering to his original interpretation.[122]

Gamow's visit to Cambridge also occurred at an opportune time for E.T.S. Walton and John Cockcroft. Only a few weeks earlier, in December 1928, Walton had suggested to Rutherford the possibility of constructing a linear accelerator,[123] and Gamow's arrival with his α-penetration calculations in hand prompted discussions, and further simple calculations, which indicated that relatively low-energy *protons* stood a reasonable chance of disintegrating light nuclei.[124] The result was that Cockcroft and Walton, with Rutherford's support, joined forces, constructed their accelerator, and ultimately succeeded in disintegrating lithium with protons in 1932.[125]

Gamow, therefore, made his personal influence felt in many different ways in Cambridge and beyond. And the influence of his, and of Gurney and Condon's, theory of α-decay penetrated deeply into nuclear physics, for it constituted the first step in establishing that the nucleus is a quantum mechanical structure, and as such it opened up numerous new and previously unanticipated avenues of research.

For Gamow personally, his theory of α-decay inaugurated his career as an independent research physicist and led to further contributions of great importance to theoretical nuclear physics. Even before leaving Copenhagen for Cambridge in January 1929 — perhaps as a result of reflecting upon the very shape of his nuclear potential with its sharp boundary between the attractive and repulsive regions — Gamow conceived the liquid-drop model of the nucleus,[126] outlined its basic features at the Royal Society discussion of February 7, 1929,[127] and soon thereafter applied it quantitatively to the problem of understanding the shape of the nuclear mass-defect curve[128] — a subject that I intend to treat in another paper. On his return to Copenhagen, he was drawn into correspondence with Houtermans which led to his making a significant contribution to Atkinson and Houtermans' theory of the building up of elements in stars.[129] One year later, in July 1930, he provided the first correct interpretation of the connection between α-particle and γ-ray emission,[130] insisting, in his inimitable way, upon signing the resulting paper on top of a mountain in Switzerland, Piz da Daint, and offering his best thanks to his companions, R. Peierls and L. Rosenfeld, "for the opportunity to work here." And the following year, he spent much of his time writing his remarkable first book, *Constitution of Atomic Nuclei and Radioactivity*,[131] in which he expressed his qualms over the behavior of the mysterious nuclear electrons by placing special symbols before and after each paragraph in which they played a role — skull-and-crossbones in the original manuscript, which Oxford University Press replaced with conservative "Lazy Ss."[132]

Gamow experienced deep changes in his personal situation as well during these years.[133] In May 1929, after almost a year in Europe, he returned to Leningrad and Russia as a hero for demonstrating to the world that Russian science — or at least one Russian scientist — was the equal of any in the world. By October he was back in Cambridge for the academic year 1929–1930 on a Rockefeller Foundation Fellowship that Rutherford had arranged for him. The following academic year, 1930–1931, having experienced no difficulty in extending his passport, he transferred to Bohr's Institute in Copenhagen. At the end of the summer of 1931, to enable him to deliver a paper in October at a conference in Rome organized by Enrico Fermi, he took a direct flight from Copenhagen to Moscow to once again renew his passport — and found the political situation dramatically changed for the worse. Ultimately, after repeated attempts, he was refused a passport, forcing his paper to be read for him in Rome by Max Delbrück.[134] Mother Russia had at last recovered her precious intellectual ore.

. But not for good. Gamow married and spent much of the next two years in planning means of escape. The opportunity came — mysteriously, for both himself and his wife — when he was allowed to attend the seventh Solvay Conference in Brussels in October 1933. Less than a year later, when Peter Kapitza returned to Moscow from Cambridge and found himself with no chance of escape whatsoever, Gamow could joke by referring to Kapitza's internment as a radiationless capture.[135] By analogy, and fortunately for physics in Gamow's adopted country, the United States, we perhaps might refer to Gamow's last two years in Russia as a transition to an excited metastable state.

Notes

It gives me great pleasure to thank the Volkswagen Foundation for financial support while carrying out this research as Visiting Professor in Germany. I also am most grateful to Dr. Ernst H. Berninger, Director of the Forschungsinstitut für die Geschichte der Naturwissenschaften und der Technik, Deutsches Museum, Munich, for providing me with office space, and for his and his colleague's warm hospitality. Figure 1 has been reproduced with the permission of *The Philosophical Magazine*; Figures 2–6 and 10–12 with the permission of the *Zeitschrift für Physik*; Figure 7 with the permission of *Nature*; and Figures 8–9 with the permission of *The Physical Review*.

1   L. Rosenfeld, "Nuclear Physics, Past and Future," in: *Nuclear Structure Study with Neutrons*, eds. M. Nève de Mévergnies, P. Van Assche, and J. Vervier (Amsterdam: North Holland, 1966), p. 483.

2   Interview with Charles Weiner, April 25, 1968, A.I.P. Center for History of Physics, New York, p. 12. Italics added.

3   H. Bethe, "Nuclear Physics B. Nuclear Dynamics, Theoretical," *Rev. Mod. Phys.* 9 (1937): 161.

4   E. Rutherford and T. Royds, "Spectrum of the Radium Emanation," *Nature* 78 (1908): 220–221; *Phil. Mag.* 16 (1908): 313–317; reprinted in: *The Collected Papers of Lord Rutherford of Nelson* (hereafter cited as *CPR*), ed. James Chadwick, Vol. 2 (London: George Allen and Unwin, 1963), pp. 70–71, 84–88.

5   E. Rutherford, "The Scattering of α and β Particles by Matter and the Structure of the Atom," *Phil. Mag.* 21 (1911): 669–688; reprinted in *CPR*, Vol. 2, pp. 238–254. See also John L. Heilbron, "The Scattering of α and β Particles and Rutherford's Atom," *Arch. Hist. Exact Sci.* 4 (1968): 247–307.

6   O. v. Baeyer, O. Hahn and Lise Meitner, "Über die β-Strahlen des aktiven Niederschlags des Thoriums," *Phys. Zeit.* 12 (1911): 273–279; "Nachweis von β-Strahlen bei Radium D," *ibid.*: 378–379; "Das Magnetische Spektrum der β-Strahlen des Thoriums," *ibid.* 13

(1912): 264–266. These results were confirmed by J. Danysz, "Sur les rayons β de la famille du radium," *Comptes rendus* 153 (1911): 339–341; *Le Radium* 9 (1912): 1–5.

7   E. Rutherford, "The Origin of β and γ Rays from Radioactive Substances," *Phil. Mag.* 24 (1912): 453–462; reprinted in: *CPR*, Vol. 2, pp. 280–287 (quote on p. 286).

8   A. van den Broek, "Infra-atomic Charge," *Nature* 92 (1913): 373.

9   E. Rutherford and H. Robinson, "Heating Effect of Radium and its Emanation," *Phil. Mag.* 25 (1913): 312–330; reprinted in: *CPR*, Vol. 2, pp. 312–327.

10 E. Rutherford, "The Structure of the Atom," *Phil. Mag.* 27 (1914): 488–498; reprinted in: *CPR*, Vol. 2, pp. 423–431.

11 J. Chadwick, "Intensitätsverteilung im magnetischen Spektrum der β-Strahlen von Radium B+C," *Ber. d. Deutsch. Phys. Gesell.* 12 (1914): 383–391.

12 R.H. Stuewer, "The Nuclear Electron Hypothesis," in: *Otto Hahn and the Rise of Nuclear Physics*, ed. William R. Shea (Dordrecht / Boston / Lancaster: D. Reidel, 1983), pp. 19–67.

13 E. Rutherford, "Collision of α Particles with Light Atoms. IV. An Anomalous Effect in Nitrogen," *Phil. Mag.* 37 (1919): 581–587; reprinted in: *CPR*, Vol. 2, pp. 585–590.

14 E. Rutherford, "Nuclear Constitution of Atoms," *Proc. Roy. Soc.* [A] 97 (1920): 374–400; reprinted in: *CPR*, Vol. 3, pp. 14–38.

15 See Stuewer, "The Nuclear Electron Hypothesis" (note 12).

16 See "Anomalous Effect" (note 13), *CPR*, p. 589.

17 E. Rutherford and J. Chadwick, "The Artificial Disintegration of Light Elements," *Phil. Mag.* 42 (1921): 809–825; reprinted in: *CPR*, Vol. 3, pp. 48–62 (figure on p. 60).

18 P.M.S. Blackett, "The Ejection of Protons from Nitrogen Nuclei, Photographed by the Wilson Method," *Proc. Roy. Soc.* [A] 107 (1925): 349–360.

19 E. Rutherford and J. Chadwick, "The Disintegration of Elements by α Particles," *Phil. Mag.* 44 (1922): 417–432; reprinted in: *CPR*, Vol. 3, pp. 67–80.

20 See his comments in: *Nuclear Physics in Retrospect: Proceedings of a Symposium on the 1930s*, ed. Roger H. Stuewer (Minneapolis: University of Minnesota Press, 1979), p. 321.

21 Karl K. Darrow, "Some Contemporary Advances in Physics — XXII Transmutation," *Bell Sys. Tech. J.* 10 (1931): 628–655; reprinted in *Bell Tel. Sys. Tech. Pub.* (Monograph B-596), 28 pp. (quote on p. 14, where the controversy is also summarized).

22 One key document in the controversy is J. Chadwick, "Observations Concerning the Artificial Disintegration of Elements," *Phil. Mag.* 2 (1926): 1056–1075.

23 See Chadwick's letters to E. Rutherford, December 9 and 12 [1927], in the Rutherford Correspondence, Cambridge University Library (hereafter RC).

24 See Chadwick's interview with Charles Weiner, April 15–21, 1969, A.I.P. Center for History of Physics, New York, pp. 61–63.

25 See for example E. Rutherford, F.A.B. Ward, and C. E. Wynn-Williams, "A New Method of Analysis of Groups of Alpha Rays. (1) The Alpha-Rays from Radium C, Thorium C, and Actinium C," *Proc. Roy. Soc.* [A] 129 (1930): 211–234; reprinted in: *CPR*, Vol. 3, pp. 225–246.

26 E. Rutherford and J. Chadwick, "Scattering of α-particles by Atomic Nuclei and the Law of Force," *Phil. Mag.* 50 (1925): 889–913; reprinted in: *CPR*, Vol. 3, pp. 143–163.

27 P. Debye and W. Hardmeier, "Anomale Zerstreuung von α-Strahlen," *Phys. Zeit.* 27 (1926): 196–199.

28 E. Rutherford, "Atomic Nuclei and their Transformations," *Proc. Phys. Soc.* 39 (1927): 359–372; reprinted in: *CPR*, Vol. 3, pp. 164–180; see especially pp. 178–179.

29 E. Rutherford, *Phil. Mag.* 4 (1927): 580–605; reprinted in: *CPR*, Vol. 3, pp. 181–202.

30 H. Geiger and J.M. Nuttall, "The Ranges of the α Particles from Various Radioactive Substances and a Relation between Range and Period of Transformation," *Phil. Mag.* 22 (1911): 613–629; "The Ranges of the α Particles from Uranium," *ibid.* 23 (1912): 439–445.

31   E. Rutherford, "Structure" (note 29), *CPR*, Vol. 3, p. 196.
32   For further biographical information see George Gamow, *My World Line: An Informal Autobiography* (New York: Viking Press, 1970) and Gamow's interview with Charles Weiner (note 2).
33   G. Gamow and D. Ivanenko, "Zur Wellentheorie der Materie," *Zeit. f. Phys.* 39 (1926): 865–868.
34   Gamow recalled (*My World Line*, note 32, p. 52) that the resulting publication, W. Prokofiew and G. Gamow, "Anomale Dispersion an den Linien der Hauptserie des Kaliums (Verhältnis der Dispersionskonstanten des roten und violetten Dubletts)," *Zeit. f. Phys.* 44 (1927): 887–892, took him completely by surprise when it appeared — Rogdestvenski had given the research to Prokofiew for completion without Gamow's knowledge.
35   See *My World Line* (note 32), pp. 52–54.
36   See note 29.
37   *My World Line*, p. 60.
38   See Ulam's "Foreword" to *My World Line*, p. ix.
39   G. Gamow, "Zur Quantentheorie des Atomkernes," *Zeit. f. Phys.* 51 (1928): 204–212.
40   D. Enskog, "Das Bohrsche Magneton und die Radioaktivität," *Zeit. f. Phys.* 45 (1927): 852–868.
41   D. Enskog, "Magnetismus und Kernbau," *Zeit. f. Phys.* 52 (1928): 203–220; "Über den Verlauf der α-Umwandlung," *ibid.* 53 (1929): 639–645.
42   Gamow, "Quantentheorie" (note 39), p. 204.
43   *Ibid.*, p. 205. See also J. R. Oppenheimer, "Three Notes on the Quantum Theory of Aperiodic Effects," *Phys. Rev.* 31 (1928): 66–81, and Lothar Nordheim, "Zur Theorie der Thermischen Emission und der Reflexion von Electronen an Metallen," *Zeit. f. Phys.* 46 (1927): 833–855.
44   Gamow, "Quantentheorie" (note 39), p. 208. The barrier width has been misprinted as e instead of l.
45   *My World Line* (note 32), pp. 60–61.
46   "Quantentheorie" (note 39), p. 212.
47   "Nuclear Physics" (note 1), p. 483.
48   M. Born, "Zur Theorie des Kernzerfalls," *Zeit. f. Phys.* 58 (1929): 306–321.
49   *My World Line* (note 32), p. 62.
50   G. Gamow and F. G. Houtermans, "Zur Quantenmechanik des radioaktiven Kerns," *Zeit. f. Phys.* 52 (1928): 496–509.
51   This expression for the decay constant, as Gregory Breit pointed out to Gamow and Houtermans by letter, contains a calculational error: the first two terms should be $\ln (4\pi m/h) + 2 \ln v$. The difference is unimportant, as it results only in slightly larger absolute values assumed for the radii r_0. See R. Atkinson and F.G. Houtermans, "Zur Quantenmechanik der α-Strahlung," *Zeit. f. Phys.* 58 (1929): 493, footnote.
52   "Quantenmechanik" (note 50), p. 509.
53   *My World Line* (note 32), pp. 63–64. It appears that in the event Gamow's stipend came from the Rask-Ørsted-Fond. See G. Gamow, "Bemerkung zur Quantentheorie des radioaktiven Zerfalls," *Zeit. f. Phys.* 53 (1929): 604.
54   These letters are in the Bohr Scientific Correspondence (hereafter BSC) in the Archive for History of Quantum Physics (hereafter AHQP). There are copies of the AHQP in the A.I.P. Center for History of Physics, New York; the American Philosophical Society Library, Philadelphia; the Bohr Institute, Copenhagen; the University of California, Berkeley; the University of Minnesota, Minneapolis; the Accademia dei XL, Rome; the Science Museum, London; and the Deutsches Museum, Munich. By October 25, 1928, Bohr could write to Joffé extolling Gamow's scientific gifts. Joffé actually did not receive

this letter, so Bohr had to write him again on December 27 enclosing a copy of his earlier letter.

55 R.W. Gurney and E.U. Condon, *Nature*, 122 (1928): 439. Gurney and Condon's concluding sentences pleased both authors. See E. U. Condon, "Tunneling — How It All Started," *Amer. J. Phys.* 46 (1978): 319–323, especially p. 320.

56 G. Gamow "The Quantum Theory of Nuclear Disintegration," *Nature* 122 (1928): 805–806.

57 R. W. Gurney and E. U. Condon, "Quantum Mechanics and Radioactive Disintegration," *Phys. Rev.* 33 (1929): 127–140.

58 *Ibid.*: 127, footnote. The occasion coincided with the dedication of the new University of Minnessota physics building. See Condon, "Tunneling" (note 55), p. 320.

59 "Tunneling" (note 55), pp. 319 and 322.

60 J.R. Oppenheimer, "On the Quantum Theory of the Autoelectric Field Currents," *Proc. Nat. Acad. Sci.* 14 (1928): 363–365.

61 R.H. Fowler and L. Nordheim, "Electron Emission in Intense Electric Fields," *Proc. Roy. Soc.* [A] 119 (1928): 173–181.

62 See N. F. Mott's obituary notice of Gurney in *Nature* 171 (1953): 910. Condon's article (note 55, pp. 321–322) makes it clear that Gurney's textbook writing after the war was to a great degree necessitated by his inability to secure a security clearance and hence a stable position, for unknown political reasons.

63 "Quantum Mechanics" (note 57), p. 130.

64 G. Wentzel, "Eine Verallgemeinerung der Quantenbedingungen für die Zwecke der Wellenmechanik," *Zeit. f. Phys.* 38 (1926): 518–529.

65 See note 40.

66 "Quantum Mechanics" (note 58).

67 *Ibid.*, pp. 137–138. Also see Condon, "Tunneling" (note 55), p. 320.

68 Gamow, "Quantum Theory" (note 57), p. 806.

69 G. Gamow, "Zur Quantentheorie der Atomzertrümmerung," *Zeit. f. Phys.* 52 (1928): 510–515 (the quote is on p. 510).

70 E. S. Bieler, "The Large-Angle Scattering of α-Particles by Light Nuclei," *Proc. Roy. Soc.* [A] 105 (1924): 434–450.

71 Gamow, "Zur Quantentheorie" (note 69), p. 513.

72 *Ibid.*, p. 514.

73 W. Bothe and H. Fränz, "Atomzertrümmerung durch α-Strahlen von Polonium," *Zeit. f. Phys.* 43 (1927): 456–465; "Atomtrümmer, reflektierte α-Teilchen und durch α-Strahlen erregte Röntgenstrahlen," *ibid.* 49 (1928): 1–26.

74 Gamow, "Zur Quantentheorie" (note 69), p. 515. Gamow cited the single offending paper, G. Kirsch and H. Pettersson, "Die Zerlegung der Elemente durch Atomzertrümmerung," *Zeit. f. Phys.* 42 (1927): 641–678.

75 Gamow, "Zur Quantentheorie" (note 69), p. 515.

76 M. von Laue, "Notiz zur Quantentheorie des Atomkerns," *Zeit. f. Phys.* 52 (1928): 726–734.

77 See note 57.

78 "Notiz" (note 76), p. 730.

79 *Ibid.*, p. 733. Von Laue cited Nernst's book, *Das Weltgebäude im Licht der Neueren Forschung* (Berlin: Springer, 1921), for the latter's hypothesis.

80 "Bemerkung" (note 53), 601–604. Gamow acknowledges von Laue's gesture in his first sentence, and the discussions with Bohr in his last.

81 G.I. Pokrowski, "Über das Herausschleudern von α-Teilchen aus Atomkernen radioaktiver Stoffe durch kurzwellige Strahlung," *Zeit. f. Phys.* 59 (1930): 427–432; Part II, *ibid.* 60 (1930): 845–849; "Über eine mögliche Wirkung kurzwelliger Strahlung auf

Atomkerne," *ibid.* 63 (1930): 561–573; "Zur Theorie der möglichen Wirkung von Strahlung auf Atomkerne," *Ann. d. Phys.* 9 (1931): 505–512.

82   H. Herszfinkiel and H. Dobrowolska, "Zu Herrn G.I. Pokrowskis Arbeiten, etc.," *Zeit. f. Phys.* 62 (1930): 432–434.

83   J. Kudar, "Bemerkung zur quantenmechanischen Deutung der Radioaktivität," *Zeit. f. Phys.* 53 (1929): 61–66; "Zur Quantenmechanik der Radioakivität," *ibid.* 53 (1929): 95–99, 134–137, and *ibid.* 54 (1929): 297–299 (Nachtrag); "Die wellenmechanische Bedingung für die Stabilität der Atomkerne," *ibid.* 57 (1929): 710–712; "Über die Verweilzeit der Korpuskeln im Gebiet der 'negativen kinetischen Energie'," *ibid.* 58 (1929): 1–2. A related criticism was entered by E. H. Kennard, "Über Potentialschwellen und radioaktiven Zerfall in der Quantenmechanik," *Phys. Zeit.* 30 (1929): 495–497.

84   J. Kudar, "Wellenmechanische Begründung der Nernstschen Hypothese von der Wiederentstehung radioaktiven Elemente," *Zeit. f. Phys.* 53 (1929): 166–167; Part II, *ibid.* 60 (1930): 262–297.

85   J. Kudar, "Die wellenmechanische Charakter des β-Zerfalls," *Zeit. f. Phys.* 57 (1929): 257–260; Parts II and III, *ibid.* 60 (1930): 168–175 and 176–180; Part IV, *ibid.*: 686–689; "Die β-Strahlung und das Energieprinzip," *ibid.* 64 (1930): 402–404; "Über die Eigenschaften der Kernelektronen," *Phys. Zeit.* 32 (1931): 34–37. For Schrödinger's evaluations of Kudar, see Schrödinger to Bohr, January 1, 1929; September 25, 1930; and April 29, 1931, BSC. For Bohr's, see Bohr to Schrödinger, May 8, 1931, BSC.

86   T. Sexl, "Zur Quantentheorie des Atomkerns," *Zeit. f. Phys.* 54 (1929): 445–448; "Zur wellenmechanischen Berechnung der radioaktiven Zerfallskonstanten," *ibid.* 56 (1929): 62–71; "Zur Theorie der bei der wellenmechanischen Behandlung des radioaktiven α-Zerfalls auftretenden Differentialgleichung," *ibid.*: 72–93.

87   R. d'E. Atkinson and F. G. Houtermans, "Zur Quantenmechanik der α-Strahlung," *Zeit. f. Phys.* 58 (1929): 478–486.

88   T. Sexl, "Zur Quantenmechanik der α-Strahlung," *Zeit. f. Phys.* 59 (1930): 579–582.

89   R. d'E. Atkinson, "Über Resonanz und Dämpfung in der Theorie des Atomkerns," *Zeit. f. Phys.* 64 (1930): 507–519, especially p. 515. F.G. Houtermans, "Neuere Arbeiten über Quantentheorie des Atomkerns," *Ergeb. d. exakt. Naturw.* 9 (1930): 123–221, especially footnote 1, p. 152.

90   T. Sexl, "Zur quantitativen Theorie der radioaktiven α-Emission," *Zeit. f. Phys.* 81 (1933): 163–177 (the quote is on p. 165). See also "Zur Theorie der Atomzertrümmerung," *ibid.* 87 (1934): 105–126; "Bericht über Fragen der Kernphysik," *Phys. Zeit.* 35 (1934): 119–141.

91   Chr. Møller, "Der Vorgang des radioactiven Zerfalls unter Berücksichtigung der Relativitätstheorie," *Zeit. f. Phys.* 55 (1929): 451–466.

92   S. Gupta, Über den radioaktiven Zerfall nach den relativistischen Wellengleichungen," *Zeit. f. Phys.* 69 (1931): 686–698.

93   See note 48 (the quote is on p. 306).

94   H.B.G. Casimir, "Bemerkung zur Gamowschen Theorie des radioaktiven Zerfalls," *Physica* 1 (1934): 193–198 (the quote is on p. 193).

95   "Neuere Arbeiten" (note 89), p. 151.

96   See note 47.

97   See C. F. von Weizsäcker, *Die Atomkerne: Grundlagen und Anwendungen ihrer Theorie* (Leipzig: Akademische Verlagsgesellschaft, 1937), pp. 93–99, especially pp. 94–95.

98   See note 89.

99   "Nuclear Physics" (note 3), p. 162. Bethe himself then went on to give a derivation "which seems about the simplest of the correct ones."

100 Interview (note 2), p. 26. We also know that Ettore Majorana, at least, wrote his 1929 doctoral thesis, "Sulla meccanica dei nuclei radioattivi," on the new theory. See E.

Amaldi, "Ettore Majorana, Man and Scientist," in *Strong and Weak Interactions — Present Problems* (New York: Academic Press, 1966), p. 17.

101 See note 40.

102 See Bohr to Fowler, December 14, 1928, BSC, where Bohr proposes Gamow's visit and also tells Fowler that Hartree and Mott will be able to describe Gamow's plans and work personally. Also see Gamow's *My World Line* (note 32), pp. 66–69, where Gamow claims that Bohr wrote to Rutherford directly. However, Gamow probably was mistaken about that, as there is no letter extant in the Rutherford correspondence from Bohr during the period in question.

103 Rutherford to Bohr, December 19, 1928, BSC. Also see Hartree to Bohr, December 21, 1928, BSC, in which Hartree says he talked to Fowler, and also to Rutherford a few days earlier about Gamow's visit.

104 The exact dates of Gamow's visit are known from a letter from Bohr to Hartree, January 5, 1929, BSC, and from Bohr to Fowler, February 14, 1929, BSC.

105 *My World Line* (note 32), p. 68.

106 Mott to Bohr, undated, no doubt February 1929, BSC.

107 *Proc. Roy. Soc.* [A] 123 (1929): 373–390. Rutherford's remarks occupy pp. 373–382.

108 K. K. Darrow, "Contemporary Advances in Physics — XXVIII. The Nucleus, Third Part," *Bell Sys. Tech. J.* 13 (1934): 580–613; reprinted in *Bell Tel. Sys. Tech. Pub.* (Monograph B-810), 48 pp. (the quote is on p. 29). Rutherford's reaction was explicitly reported in Chadwick's interview (note 24), p. 51.

109 E. Rutherford, J. Chadwick, and C. D. Ellis, *Radiations from Radioactive Substances* (New York: Macmillan and Cambridge: Cambridge University Press, 1930), pp. 326–333.

110 *Ibid.*, p. 327. We know of Chadwick's authorship of these remarks from his interview (note 24), p. 49.

111 Rutherford's model was also insightfully criticized earlier by G. Gentile, "Sulla teoria dei satelliti de Rutherford," *Atti della Reale Accad. Naz. dei Lincei* 7 (1928): 346–349. E. Segrè recalls (private communication, October 1979) that Gentile often visited Rome, where he conveyed his opposition to Fermi's group.

112 See Mott to Bohr, September 16, 1930, BSC.

113 See Fowler to Bohr, April 9, 1929, BSC. Also see Fowler and A. H. Wilson, "A Detailed Study of the 'Radioactive Decay' of, and the Penetration of α-Particles into, a Simplified One-Dimensional Nucleus," *Proc. Roy. Soc.* [A] 124 (1929): 493–501.

114 R. W. Gurney, "Nuclear Levels and Artificial Disintegration," *Nature* 123 (1929): 565. Condon later pointed out that, to his chagrin, he talked Gurney out of this idea in Princeton and hence delayed its publication, but Gurney persevered and published it from Tokyo when he was "no longer subject to my bad influence." See "Tunneling" (note 55), p. 321.

115 "Detailed Study" (note 113), p. 501.

116 J. Chadwick, and G. Gamow, "Artificial Disintegration by α-Particles," *Nature* 126 (1930): 54–55.

117 R. d'E. Atkinson, "Über Resonanz und Dämpfung in der Theorie des Atomkerns," *Zeit. f. Phys.* 64 (1930): 507–519.

118 G. Hoffmann and H. Pose, "Nachweis von Atomtrümmern durch Messung eines einzelnen H-Strahls," *Zeit. f. Phys.* 56 (1929): 405–415; H. Pose, "Messungen von Atomtrümmern aus Aluminium, Beryllium, Eisen, und Kohlenstoff nach der Rückwärtsmethode," *ibid.* 60 (1930): 156–167.

119 H. Pose, "Über die diskreten Reichweitengruppen der H-Teilchen aus Aluminum. I. Abhängigkeit der Ausbeute und Energie der H-Teilchen von der Primärenergie," *Zeit. f. Phys.* 64 (1930): 1–21.

120 See for example H. Pose, "Über Richtungsverteilung der von Polonium-α-Strahlen aus

Aluminum ausgelösten H-Teilchen," *Phys. Zeit.* 31 (1930): 943–945; J. Chadwick, J.E.R. Constable, and E. C. Pollard, "Artificial Disintegration by α-Particles," *Proc. Roy. Soc.* [A] 130 (1931): 463–489; K. Diebner and H. Pose, "Über die Resonanzeindringung von α-Teilchen in den Aluminumkern," *Zeit. f. Phys.* 75 (1932): 753–762. A summary is provided in M.A. Tuve, "The Atomic Nucleus and High Voltages," *J. Franklin Inst.* 216 (1933): 1–38, especially p. 20.

121   W. Bothe and H. Becker, "Künstliche Erregung von Kern-γ-Strahlen," *Zeit. f. Phys.* 66 (1930): 289–310.

122   W. Bothe, "α-Strahlen, künstliche Kernumwandlung und -Anregung, Isotope," in: *Convegno di Fisica Nucleare Ottobre 1931–IX* (Rome: Reale Academia d'Italia, 1932–X), pp. 83–106.

123   See E.T.S. Walton to E.M. McMillan, April 11, 1977, in: *Nuclear Physics in Retrospect* (note 20), pp. 141–142.

124   Gamow, *My World Line* (note 32), p. 83.

125   J. Cockcroft and E.T.S. Walton, "Experiments with High Velocity Positive Ions. II. — The Disintegration of Elements by High Velocity Protons," *Proc. Roy. Soc.* [A] 137 (1932): 229–242.

126   See Gamow to Bohr, January 6, 1929, BSC, where Gamow reports that en route to Cambridge he stopped off in Leiden, where P. Ehrenfest was greatly interested in his "Tröpfchenmodell." Gamow was partly influenced by recent studies of Guido Beck. See for example his paper "Über die Systematik der Isotopen. II.," *Zeit. f. Phys.* 50 (1928): 548–554.

127   See note 107. Gamow's remarks are on pp. 386–387. See also "Über die Struktur des Atomkernes," *Phys. Zeit.* 30 (1929): 717–720.

128   G. Gamow, "Mass Defect Curve and Nuclear Constitution," *Proc. Roy Soc.* [A] 126 (1930): 632–644. Gamow's "Tröpfchenmodell" was widely discussed at the time; in particular it was picked up by C. F. von Weizsäcker, "Zur Theorie der Kernmassen," *Zeit. f. Phys.* 96 (1935): 431–458.

129   R.d'E. Atkinson and F.G. Houtermans, "Zur Frage der Aufbaumöglichkeit der Elemente in Sternen," *Zeit. f. Phys.* 54 (1929): 656–665; M. Delbrück and G. Gamow, "Übergangswahrscheinlichkeiten von angeregten Kernen," *ibid.* 72 (1931): 492–499.

130   G. Gamow, "Fine Structure of α-Rays," *Nature* 126 (1930): 397. The famous photograph commemorating the signing is reproduced in *My World Line* (note 32), p. 87, with, however, an incorrect caption — the other person on the photo is not Landau but Rosenfeld; Peierls took the picture. Also, for an earlier related article see "Successive α-Transformations," *Nature* 123 (1929): 606, and for a later one see "Über die Theorie des radioaktiven α-Zerfalls, der Kernzertrümmerung und die Anregung durch α-Strahlen," *Phys. Zeit.* 32 (1931): 651–655.

131   G. Gamow, *Constitution of Atomic Nuclei and Radioactivity* (Oxford: Clarendon Press, 1931). The book is dedicated "To the Cavendish Laboratory, Cambridge" and the Preface was signed in Copenhagen on May 1, 1931. Gamow had been much concerned with the behavior of the nuclear electrons for years, but especially following O. Klein's discovery of the so-called Klein paradox in Copenhagen at the end of 1928. See O. Klein, "Die Reflexion von Elektronen an einem Potentialsprung nach der relativistischen Dynamik von Dirac," *Zeit. f. Phys.* 53 (1929): 157–165.

132   See his interview (note 2), p. 34.

133   See *My World Line* (note 32), Chapters 3–6, pp. 55–133.

134   G. Gamow, "Quantum Theory of Nuclear Structure," in: *Convegno* (note 122), pp. 65–81.

135   See for example his letters to Bohr, January 20, 1935, BSC, and to Rutherford, February 5, 1935, RC.

On Gamow's Theory of Alpha-Decay
A Comment

HARRY J. LIPKIN

I first learned about Gamow's theory of alpha decay when I was a student at Princeton in 1947. To us Gamow was mainly the colorful character who had written the delightful Mr. Tompkins books and whose sense of humor was manifest in all his scientific work. His book on nuclear physics was one of the first texts available on the subject. The first edition contained a joke which many readers missed. He derived the radial Schrödinger equation, eq. (4) of Stuewer's lecture, in the usual manner by writing the equation first in cartesian coordinates in three dimensions and then transforming the equation to polar coordinates. However, the equation in polar coordinates was completely wrong. It was written in a way which looked correct to the casual reader who never checks the details and always jumps to the radial equation. But careful scrutiny revealed that it was full of mistakes. Whether Gamow simply wrote down something that looked right without checking it, or whether this was an intentional joke is not clear. But most readers of the book never noticed it because the radial equation and all the conclusions were correct.

In one of his books Gamow gives another description of how he happened to discover the theory of alpha decay in Göttingen. Göttingen, he wrote, was a sleepy town with only two movie theaters, and he discovered the theory of alpha decay because there was nothing else to do in the evenings. We

E. Ullmann-Margalit (ed.), The Kaleidoscope of Science, 187–192.

immediately noted that Princeton in 1947 was also a sleepy town with only two movie theaters and nothing else to do in the evenings. It was thus obvious why the same theory of alpha decay was simultaneously and independently discovered in Göttingen by Gamow and in Princeton by Condon and Gurney.

Professor Stuewer has given a very thorough account of the development of the basic ideas of tunneling and barrier penetration in quantum mechanics. I shall add to this a description of recent developments which have brought the problems of tunneling and barrier penetration back again as an exciting open area of frontier research.

The treatment of alpha decay by Gamow is easily extended to all problems of barrier penetration which are effectively one-dimensional, e.g. they can be reduced to a one-dimensional radial equation by separating out other degrees of freedom. But a two-dimensional problem already introduces qualitatively new difficulties. Consider the simple problem of tunneling through a mountain range on the surface of the Earth. A particle on one side of the mountains has a quantum mechanical probability of leaking through this barrier and coming out on the other side. Of course this probability is negligibly small for macroscopic mountains, but one can imagine a two-dimensional problem on the atomic scale, analogous to the simple barrier-penetration problems considered first by Gamow and now found in every quantum mechanics textbook, but different in having a highly irregular barrier like a miniature mountain range with no simple symmetries. The question arises: "Where should one put the tunnel?" It is impossible to define Gamow's barrier-penetration integral appearing in Stuewer's eq. (6) without defining some kind of path in the two-dimensional space under the mountains.

The problem becomes even more complicated in multidimensional problems arising in multiparticle dynamics and in quantum field theory, where there are many degrees of freedom. Consider, for example, the spontaneous fission of a heavy nucleus. The particles in the nucleus are initially in a state which would be stable in classical mechanics. The state is metastable in quantum mechanics because there are many available states of lower energy for this system of particles, in which the nucleus has split into two fragments. The nucleus cannot split classically because it cannot reach these lower energy states by any classical motion which conserves energy. In quantum mechanics it can tunnel through this barrier. But where is the tunnel in this multidimensional space? And how can one calculate the relative probabilities of the fission of plutonium into different kinds of fragments like barium, cesium, xenon, etc.? In conventional neutron-induced fission above the barrier, the equations of motion define a path or mechanism for the reaction in which the spherical nucleus gradually deforms to an ellipsoidal shape and finally breaks into two pieces. But in spontaneous fission there is no classical path allowed and no way to follow the detailed stages of the breakup of the nucleus. The transition proceeds somehow through the classically forbidden region in the multidimensional space. One needs to define some kind of "path" for the tunnel by which the transition occurs in order to define the Gamow penetration integral. Since there are many possible final states, there must be many tunnels, each with a different barrier-penetration factor.

A similar problem arises in the new non-Abelian quantum gauge field theories, which are now in vogue for describing the weak and strong interactions of particle physics. Cases arise where there are infinite sets of solutions for the ground state, i.e. the vacuum in the field theory, with barriers between them. The true ground state, like the true ground state in a double-well potential, will have a lower energy than the single-well solution, by an amount which depends upon the barrier-penetration factor. Here again is a barrier in a multidimensional problem, where Gamow's formula cannot be simply applied without some procedure to define the correct path in the multidimensional space.

One approach for finding the tunnel through multidimensional barriers is called the "instanton." It is based on the properties of the equations of motion, both in classical and quantum physics, when the time is analytically continued into the complex plane and imaginary values of time are used. Consider, for example, Newton's equation of motion for a particle in a one-dimensional external potential,

$$m\{d^2x/dt^2\} = -dV/dx \tag{1}$$

This equation can be rewritten

$$m\{d^2x/d(it)^2\} = -d(-V)/dx \tag{2}$$

If we now consider the equation of motion in terms of the new imaginary time variable "it", we see that it represents the motion of the particle in a potential $-V(x)$. If $V(x)$ is a double-well potential, as shown in Figure 1a, then $-V(x)$ is a double barrier, as shown in Figure 1b.

Classically, if a particle is moving with a small amplitude of oscillation about either of the two equilibrium positions of the double well of Figure 1a, it remains there forever and never gets to the other side. In quantum mechanics, however, there is tunneling through the barrier. The ground state

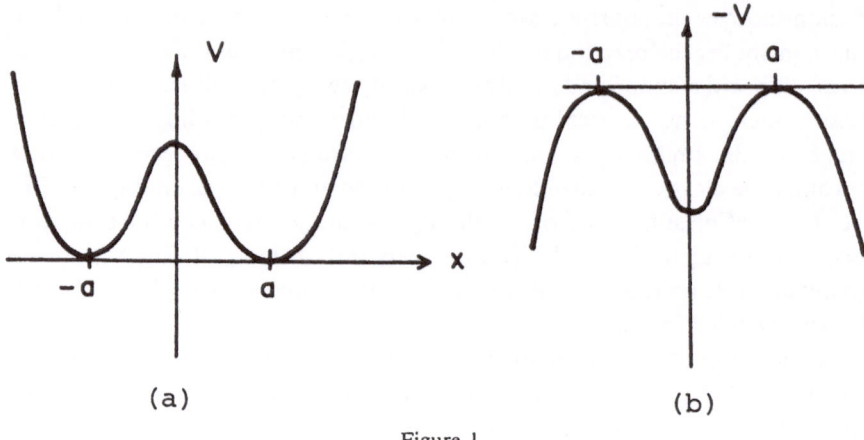

Figure 1

and first excited states of this system are the symmetric and antisymmetric combinations of the solutions for the ground state in each well, and the splitting between the two energy levels is just given by a Gamow barrier-penetration integral. In the imaginary time version of this problem with the potential of Figure 1b, the barrier has become a well and there are classical solutions traversing it. A particle can start at $x=-a$, roll down the well classically and emerge at $x=a$. This particular solution, shown in Figure 2, is called the "instanton." It has been shown that by analyzing the properties in

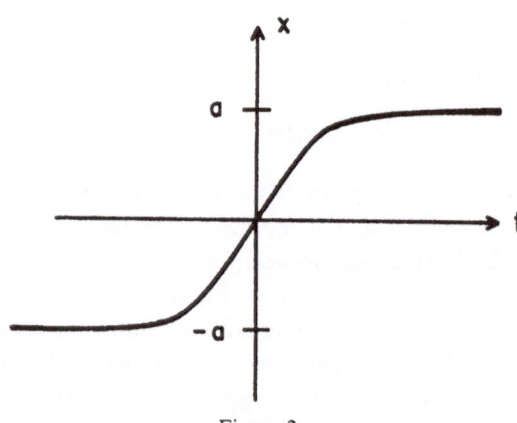

Figure 2

imaginary time of this instanton solution which traverses the region of space forbidden for the problem in real time, the same barrier-penetration integral appears, and it is possible to calulate the barrier-penetration factor.

In one dimension there is no advantage to using the instanton solution to calculate the tunneling through the barrier. But in multidimensional problems, where the simple methods are not easily applicable, it is possible to write the equations of motion in imaginary time and look for instanton solutions. These orbits in imaginary time help to show "where the tunnel is" in real time and enable the calculation of the barrier-penetration factor.

This approach is easily extended to the multidimensional case. For example, in the case of the mountain range in two dimensions with an equilibrium position on each side, the imaginary time problem changes the two equilibrium positions into two peaks, and the mountain range into a canyon. The solutions of the equations of motion for a particle which begins its motion at one equilibrium point and rolls down the canyon and up to the other equilibrium point provide the necessary information for the solution of the barrier-penetration problem in real time in quantum mechanics. These instanton solutions show "where the tunnel should be" in the barrier problem and how to calculate the Gamow penetration factor. The exact formulation of the problem is described in detail in Sidney Coleman's 1977 Erice lectures, along with the applications in quantum field theory.[1]

This instanton formulation has been used by Levit, Negele and Paltiel to treat spontaneous fission.[2] Here the nucleus is originally in a state of classical equilibrium, with a barrier between it and the state where it is broken into two fragments with the same energy. Outside the barrier there is a repulsive **Coulomb potential which drives the fragments apart.** The use of the imaginary time converts the potential barrier to a canyon through which the **system can propagate, while the repulsive Coulomb potential outside now** becomes an attractive well which prevents the fragments from escaping from one another. The motion of the system in imaginary time includes orbits in which the nucleus begins in the equilibrium state of a single nucleus and then rolls down the canyon. The system thus evolves into two fragments moving apart in an attractive well and losing kinetic energy until their motion is reversed and they bounce back to the original state. The study of the evolution of the system for this "bounce" orbit in imaginary time then provides the information of how the nucleus breaks up under the barrier, i.e. where the tunnels are in the multidimensional space. This can then be used to calculate the Gamow barrier-penetration factor and the spontaneous fission lifetime for the quantum-mechanical system in real time.

At present the real physical problems both in many-body physics and in quantum field theory are too complicated to be solved by present techniques, and the imaginary time approach is being used with various approximation methods and simplified models to treat barrier-penetration phenomena in systems with many degrees of freedom. The exciting field of barrier penetration opened by Gamow thus has not only solved the problem of alpha decay and other simple penetration problems which are essentially one-dimensional. It has also opened the way to new directions in research which are still active today.

Notes

1   S. Coleman, "The Uses of Instantons," in: Proceedings of the 1977 International School of Subnuclear Physics, "Ettore Majorana," Erice, Italy.
2   S. Levit, J. W. Negele, and Z. Paltiel, *Phys. Rev.* C22 (1980): 1979.

The Group Construction of Scientific Knowledge: Gentlemen-Specialists and the Devonian Controversy

MARTIN J. S. RUDWICK

Introduction

Several years ago Sir Peter Medawar urged analysts of science to study in detail "what scientists *do*" in the course of their research. As a distinguished practicing scientist, Medawar himself was well aware that what scientists *say* about their activity can never be taken at face value. Instead he urged that "only unstudied evidence will do — and that means listening at a keyhole."[1] In fact, however, the few analysts of science who have followed this prescriptive suggestion effectively have preferred to establish themselves as participant observers on the inner side of the keyhole, to penetrate into the laboratory itself and to watch the activities of scientists from the perspective of the ethnographer or anthropologist.[2] Valuable and provocative as such studies are, however, I think their authors have underestimated the extent to which historical studies can also contribute to a better understanding of scientific practice.[3] Clearly it is desirable in any case that we should trace how the practice of research has changed over the *longue durée* of the history of science, in conjunction with changing social and cognitive circumstances. But quite apart from that, there are also ways in which much more 'fine-grained' historical studies may give us better access than research on modern science can, to the "unstudied evidence" that Medawar saw was needed; and

193

E. Ullmann-Margalit (ed.), The Kaleidoscope of Science, 193–217.
© *1986 by D. Reidel Publsihing Company.*

they may avoid the observational or interpretative 'distance' that was implied in his metaphor of the keyhole.

Even for contemporary science, there is an ever-present danger of interpreting the activities of scientists and the processes of research practice with the benefit of hindsight. Realizing this, some cognitive sociologists of science have deliberately chosen to study problems or controversies that are as yet unresolved, in order to avoid any possible use of hindsight, by the scientists they interview or by themselves.[4] But there is another way to avoid the same danger, and it is a way that allows the historian to make a non-retrospective analysis even of problems that have long since been resolved. The method is simply to reconstruct the chosen episode with the strictest attention to that scorned and neglected component of historical practice — precise chronology. One must think oneself back into the lives of the historical actors *as their research proceeded*: not just in a general sense of getting inside the skin of their 'world-view' or even the contemporary state of their particular discipline, but in the far more specific sense of reliving what they did and said and wrote and argued about, week by week and month by month.

At present, most of the best examples of such 'fine-grained' reconstructions are focused on the work of some particular outstanding individual whose *Nachlass* is rich and complete enough to allow a detailed reconstruction of the development of his or her research. An outstanding example would be the fine body of recent research on Charles Darwin's early work on a theory of evolution.[5] Yet even here, it is only recently that historians have become aware of the dangers of retrospective analysis; only recently have some of them begun to treat each phase of Darwin's work as a cognitive entity in its own right, and not just as a building-block added to a cumulative structure, or as an almost inevitable step on Darwin's path to a finally successful theory.[6] It is at least arguable, however, that such examples of 'fine-grained' reconstruction, however brilliantly carried out, are seriously atypical, precisely because they are focused on individual scientists who were so outstandingly original that their work proceeded in relative isolation from other scientists.

Without falling into the opposite and currently fashionable trap of scientific egalitarianism, which assumes that Nobel prize-winners are no more worthy of attention than run-of-the-mill Ph.D.'s, we can surely allow that much, if not most research in science, at least in the past two centuries, involved an interactive *group* or network of specialists.[7] Yet we still lack any substantial body of examples to show precisely how, in 'fine-grained' chronological detail, new knowledge is built up out of these group interactions. Material for such examples may not be as rare as historians of science tend to assume. The contingencies of certain places and periods in science, and certain social configurations of scientists, may have preserved unexpectedly rich evidence of the social processes by which particular pieces of claimed scientific knowledge were constructed and validated. These episodes may or may not turn out to be typical of science as a whole. The very circumstances which occasioned the exceptional preservation of documentary evidence may bear witness to their atypicality. But unless we try to assemble some examples of this kind, we shall not even discover the parameters of possible variation in the historical practice of scientific research.

Background and Setting

In this paper I shall give a brief progress report on my current work of writing a 'fine-grained' and nonretrospective narrative and analysis of one such episode.[8] The period is that extraordinarily fertile time for many of the natural sciences, the second quarter of the nineteenth century. The place is mainly England and particularly London, but with important extensions to Paris, to the rest of Europe, and, eventually, to the rest of the world. Combining the place and the period, it is not surprising that the social setting is that of 'gentlemen of science'.[9] The main historical actors were *gentlemen-specialists*: gentlemen both by their social class, and by their possession of resources that enabled them to carry out substantial scientific research without the need of paid employment for it; specialists by virtue of the fact that even the most polymathic of them concentrated their main efforts in one or a few specialized branches of natural knowledge.[10] The science chosen in this instance was geology, which was then experiencing its first and greatest boom in conceptual innovation, empirical expansion, and public approval and interest.[11] The episode is one that was widely known and discussed at the time in scientific circles, though it has since dropped into obscurity; but this is a positive advantage for the historian, because the story can be told without the readers knowing beforehand who were the goodies and who the baddies, or how the plot was to end.

I first became aware of the importance of this episode many years ago, while I was sorting through the scientific correspondence — at that time still privately owned — of one prominent geologist of the period.[12] For I came across a bundle of letters, apart from the rest, tied up with red tape and labeled 'Great Devonian Controversy'. Reading these letters made me aware that here were the traces of a controversy that had evidently been central to geology for almost a decade. It had raised questions at the heart of contemporary geological practice, and it had had far-reaching implications for the careers of some of those involved. But above all, these letters revealed at once the kind of vigorous informal argument, combative and persuasive by turns, that I was then experiencing at first hand as a practicing scientist. Yet I knew also from that experience that in the mid-twentieth century such informal arguments vanish almost entirely into thin air above the coffee cups at scientific conferences and similar occasions, leaving only the most distorted and misleading traces in formal papers and the unreliable memories of participants. It became clear to me, therefore, that the Devonian controversy might be an unusually favorable strategic site for an analysis of "what scientists *do*," or at least what they *did* a century and a half ago in one specific social setting. So in the interstices of other research I began tracking down the other sides of this correspondence in other archives. By an informal process, analogous to the sociologists' technique of snowball sampling, the network of those who turned out to have been involved in the Devonian controversy grew slowly outwards from where I had first happened to find it, until it came to comprise — in minor if not major roles — most of the leading geologists of the time in Britain and many of those on the Continent too.

It was not only leading geologists, however, who played roles of various kinds in the Devonian controversy. Any adequate narrative and analysis of this episode must also attend to the roles of many lesser figures too. In place of the customary dichotomy between active scientific performers and their relatively passive audience, with its implicit definition of a strong-boundaried scientific community, we need a mental image of the social and cognitive 'topography' of scientists that allows for much weaker boundaries and for many kinds of contingent variation in different sciences at various periods. For geology at the time of the Devonian controversy, a topography of concentric weak-boundaried *zones of competence* is appropriate (Figure 1).[13] Such zones, and the individuals who populated them, can be recovered from the historical record by analyzing the ways in which specific individuals can be seen to have treated the work of others. This analysis recovers a tacit pecking order or gradient of competence in geology — the competence being of course that ascribed by the geologist at the time to themselves and others. Even where they argued vehemently with each other, they implicitly acknowledged that their opponents' arguments needed to be taken seriously. So there was a high degree of tacit consensus about the form of this invisible topography: those in zones of lesser competence generally accepted that their proper place was there, even if they had ambitions to climb higher.[14]

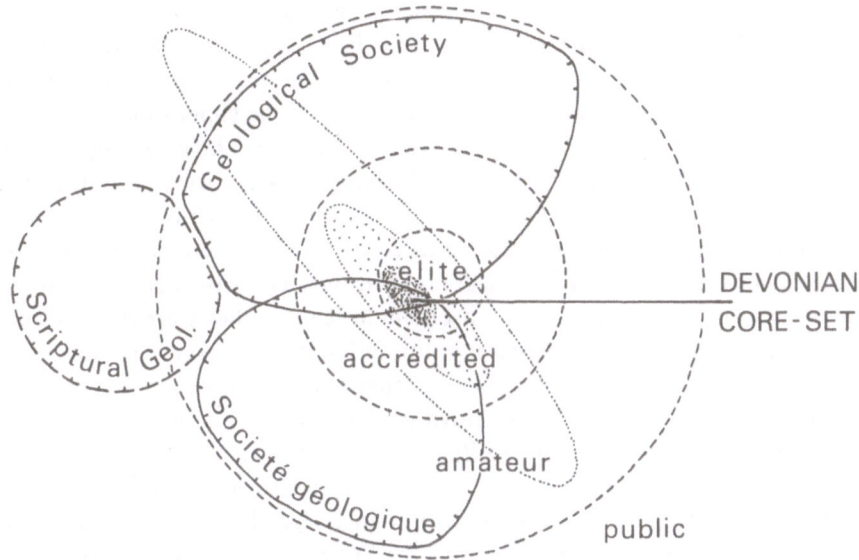

Figure 1. The social and cognitive topography of geology in the 1830s, and of the Devonian controversy, drawn as a Venn diagram. This shows (a) the three weak-boundaried concentric zones of ascribed competence in geology (elite, accredited, amateur) surrounded by the general public; (b) the analogous zones of relative involvement in the Devonian controversy (the 'core-set' of major involvement is densely stippled; the surrounding 'matrix' of moderate or slight involvement is stippled more lightly or left blank); and (c) the strong-boundaried major geological societies (Geological Society of London, Societé géologique de la France) and the marginal "scriptural geologists."

Geologists in the zone I have termed the 'elite' deemed themselves competent to put forward theoretical interpretations of the Earth and its history at the highest level of generality. 'Accredited' geologists, by contrast, were deemed competent to give reliable accounts of the geology of the regions or strata or other phenomena of which they had first-hand knowledge, but not to give high-level interpretations beyond those limits. 'Amateurs', in the sense I use the term here, were deemed competent to make reliable local observations and to assemble reliable collections of specimens, but not to offer even local interpretations of their significance. Members of the general public were not even deemed competent to make reliable observations: their reports were treated with scepticism, unless and until they had been checked by someone with at least 'amateur' status.[15]

An episode like the Devonian controversy was a *focal problem* or temporary 'hot-spot' that disturbed the otherwise even tenor of much routine research in the science.[16] As such, it was of central importance to a certain subset of the elite geologists, and it also engaged the attention of subsets of the accredited and amateur geologists in major or minor ways. To represent this, a second-order 'topography' must be superimposed on the first, making it a complex Venn diagram (Figure 1). Those elite geologists who were involved in the controversy as a major part of their current work, and who were familiar with all the relevant arguments as it proceeded, were also those who — by virtue of their elite status — were deemed competent to pronounce on its widest theoretical implications. These men constituted the 'core-set' for this particular focal problem in science; they were the small set of those through whose changing opinions the controversy was ultimately deemed to have been settled.[17] Beyond that core-set, however, was a wider circle or matrix of those who contributed to the debate in important but lesser ways. Some of them were elite geologists who might simultaneously be in the core-set that was debating some other focal problem in the science.[18] Others were accredited geologists with highly regarded local or specialized expertise that was deemed relevant to the Devonian problem. Others again were merely amateur geologists who assembled reliable local collections of specimens, which were then offered to those in higher zones of competence for evaluation and interpretation. Finally there were some who did not even enjoy amateur status — notably quarrymen and miners from far below the gentlemanly social class — whose collections, made for payment, were nonetheless accepted as important once their authenticity had been checked.

These two graded social topographies — the one for the Devonian focal problem being in a complex manner a subset of the one for geology as a whole — must also both be regarded as cognitive in character, since they were tacitly defined by the ways in which claimed knowledge or information was treated according to its point of origin and in its subsequent transmission across the topography. Of course, no visual representation of such topographies can adequately depict the subtleties of the real social and cognitive interactions that sustained them. But the value of such a diagram is that it encourages the historian to attend to the structure of the *whole* network of information-flow through which new knowledge was constructed.[19] Certainly in the case of the Devonian controversy, any analysis would be seriously defective if it ignored the effects on the debate of empirical material derived from even the lowest zones of ascribed competence. Specimens collected by amateurs or quarrymen with little or no understanding of the higher theoretical issues were nonetheless a vitally important empirical input into the debate.

My diagram, however inadequate, does at least do rough justice to the social shape of the whole of geology as it was practiced internationally at the time of the Devonian controversy, since there was substantial communication and mutual evaluation between the main centers of research in different countries. The formal institutions serving geology at this time — the sister societies in London and Paris were by far the most important — can be mapped as a further element onto the informal tacit topography, making it a still more complex Venn diagram (Figure 1).[20] If this topography represents the entire social 'field' of geology, then the Geological Society of London can be identified as the most important arena for the social dramas that formed the high points of the Devonian controversy.[21] It was the oldest society anywhere to be devoted to geology; but it was also by common consent one of the liveliest arenas for debate in *any* science at that period. Alone among learned societies, it allowed and even encouraged argumentative discussion after the reading of scientific papers. Even quite acrimonious argument was in practice tolerated, because it took place among gentlemanly social equals, and no account of these private discussions was allowed to be published.

The importance of this social convention became apparent when, at one crucial phase of the Devonian controversy, similar arguments were aired in a very different arena. At the annual meetings of the newly founded British Association for the Advancement of Science, the 'Section' devoted to geology was organized by much the same group of men who ran the London society. But meetings of the Association were open to a much wider social range of participants (including, grudgingly, the female sex).[22] So there was great concern among geologists when divergent opinions about the correct geological structure of Devonshire were discussed in this socially less exclusive arena, because one leading geologist (Roderick Murchison) publicly impugned the scientific competence of another (Henry De la Beche) and thereby imperiled the latter's career prospects.[23] But this was in fact almost the only point at which macrosocial factors impinged on the Devonian controversy. These two prominent geologists epitomized two alternative futures for their science within British society: the one (De la Beche) seeking financial support from the government and anticipating a more professionalized future; the other (Murchison) looking to aristocratic circles for social approval and a more diffuse sense of legitimacy for the life of science.

Undoubtedly, this contrast contributed to the acrimony of their argument in the Devonian controversy. Yet there is no adequate evidence to support an interpretation of the controversy as a whole in terms of the divergent social interests they represented. There was no discernible pattern of social alignment within the controversy that would correlate with such interests; indeed the very fluidity of the cognitive alignments within the controversy, when it is traced in detail, makes any such interpretation highly implausible. While social interests, along with ideological conflicts, probably did operate constitutively in certain other controversies in nineteenth-century science, in the Devonian controversy their cognitive effect was marginal. So was that of the occasional charges of scandalous conduct or attempted fraud. But this does not mean that the cognitive processes involved in the Devonian controversy were therefore nonsocial. On the contrary, a 'fine-grained' analysis shows that the normal processes of scientific practice were *social* processes through and through, in the Devonian controversy as in any other example of the construction of scientific knowledge.[24]

Private and Public Science

The historian of science who studies any period before the twentieth century is deprived of the sociologist's classic form of evidence, the interview. But historians who have had to deal with the written reminiscences of earlier generations of scientists may well feel that this deprivation is less than disastrous. For the recollections of participants, years or decades after the events being recalled, are notoriously and systematically unreliable.[25] The historian of science is on safer ground if forced by the nature of the record to focus attention on material that is strictly contemporary with the events being reconstructed and analyzed. In fact, to describe the forms of documentary evidence is, in effect, to outline simultaneously the shape of scientific practice. The historian may have no unmediated access to the thoughts of the scientists being studied; but the documents certainly give mediated access to the activities that constituted their practice.

Instead of the customary sharp dichotomy between published and unpublished material, it is more useful to review the range of documentary evidence in terms of a continuum stretching from the most private to the most public forms of activity and their corresponding records.[26] Taking the Devonian controversy as a concrete example, and beginning at the private end of the continuum, the historian has access firstly to field notebooks, which constitute a trace of the day-by-day activities of at least the more active individuals, during their periods of most intensive '*intra-*personal' interaction with the natural objects that were accepted as the empirical basis for all theoretical arguments. Geology in the early nineteenth century was above all a field science; the empirical focus of geological practice was the field excursion, far more than the laboratory or even the museum. Few geologists kept daily records of their research, except during periods of field work. But what field notebooks lack in continuity, in comparison with laboratory notebooks in the experimental sciences, they may amply compensate through vivid immediacy.[27] They were often compiled by a solitary geologist or simultaneously by a pair of colleagues, during prolonged periods of study in romantic rural scenery. Such individuals were far from home, and temporarily suspended in 'liminoid' isolation away from the ordinary pressures of the larger disciplinary community.[28] In these notebooks, intended only for the compiler's own eyes, the mundane details of observation are found inseparably linked to their spontaneous provisional interpretation. And while deciphering and reconstructing the daily course of field work, the historian's perseverance is occasionally rewarded by reading an interpretative note which gives a sudden insight of startling transparency into the line of reasoning that the individual was following on the most private level at that moment.

But notebooks give no specially privileged access to the thought processes of the individual. They record one relatively private aspect of the whole structure of scientific practice, but they also reveal the integration of solitary reasoning into a larger fabric of social interaction. Interpretative notes are often explicit rehearsals of arguments to be used against real or imagined critics. Even the most innocently 'factual' of observations can often be seen from their context to have been guided by theoretical objectives, such as the need to search for new empirical evidence to reinforce the compiler's own position or to undermine that of his critics.

Closely linked to the evidence of notebooks, but opening out away from the private end of the continuum, is the evidence of correspondence, the trace of generally dyadic or *inter*-personal interactions. In the case of the Devonian controversy, this evidence is quite exceptionally rich. This may be due partly to the perceived importance of the controversy at the time, which may have led the main participants to preserve the relevant letters with particular care.[29] Partly it is also the result of various contingent circumstances which caused those participants to be geographically separated for much of the time, although they knew each other well enough to correspond on easy and informal terms, even when vehemently criticizing each other's opinions.[30] More generally, they were not cursed with the telephone; and the postal service — particularly within Britain — was much more efficient than it is today.[31] And they belonged to a society — or more accurately, to a social stratum within that society — in which a spontaneous and vigorous style of letter-writing was a routine accomplishment. All these circumstances combine to provide, in the case of the Devonian controversy, an exceptionally well-preserved record of a network of interactions. This network linked individuals situated in all parts of the social-cognitive topography of the science, from elite to amateur geologists; and it can be reconstituted at a level of temporal resolution, which at times becomes as fine-grained as several times a week.

Like field or laboratory notebooks, scientific correspondence gives the historian no straightforward or unproblematic trace of individual thoughts or opinions. Even more clearly than notebooks, letters were at this period an indispensable medium for the continuing process of argument and persuasion underlying the construction of knowledge. In these exchanges of letters — sometimes dashed off in haste, sometimes evidently composed with deliberate care — are preserved the traces of complex moves that were at one and the same time both social and cognitive. If read in a strictly contemporary context, even the most apparently straightforward factual reports can be seen to have been directed by the writer toward altering the opinions of the recipient: consolidating one alliance, detaching the recipient from another, enhancing the writer's own credibility, undermining the credibility of one of his critics. Nothing is quite what it appears on the surface: the most significant remarks may be introduced with an apparently casual "By the way..." or relegated to a "P.S." Needless to say, such 'readings' by the historian are themselves never uniquely definitive; but they can and do receive cumulative corroboration, the more fully the individual letter is embedded in its immediate context of contemporary argument.[32] It is here, perhaps more than at any other point in the task of reconstruction, that precise chronology becomes essential; for even a single letter, misdated perhaps by careless deciphering of a smudged postmark, can throw the meaning of many other letters into confusion.

The informality of these exchanges by letter — even when they were acrimonious — was closely related to the character of the main arena in which the drama of the Devonian controversy was played out. As already mentioned, the Geological Society of London was renowned at the time for the vigor of its argumentative discussions. These discussions, and the formally presented papers that occasioned them, constitute a further stage along the continuum from private to public science; but, as noted earlier, they were significantly less than fully public events. Formal summaries of the papers were published promptly in the Society's *Proceedings*, circulated among the membership, and quite widely reprinted or abstracted in general scientific periodicals in Britain and abroad. But, like modern scientific papers, they are often highly misleading in that they systematically omitted or concealed much that was speculative or controversial. Reports of the reading of such papers, as recorded in letters to absent members of the Society, are often more reliable sources, at least to supplement the printed record, even if the historian has to make allowance for the known biases of the writer. And such private and informal reports are the *only* source from which the course of the famous discussions themselves can be reconstructed. Nonetheless, at least for some occasions the discussions can be reconstructed with some confidence, particularly when independent accounts of what was said can be checked against each other. Though often faint and obscure, such accounts of discussions are the traces not only of face-to-face interactions across the floor of the Society's meeting room, but also of those more subtle '*trans*-personal' interactions by which it seemed occasionally that the meeting as a whole had reached some kind of provisional collective opinion transcending those of the individual geologists present.[33]

Finally, at the most fully public end of the continuum, relevant papers provided the occasion for almost annual discussions of the controversy at the meetings of the British Association, open to all who could pay the fee, and fully reported in the general cultural weeklies (particularly the *Athenaeum*) and the scientific periodicals.[34] On other occasions, when priority had to be established in a hurry, or when the controversy became more personally acrimonious than the guardians of the Geological Society's gentlemanly norms would tolerate, the usual procedures were simply bypassed, and 'letters to the editor' were given rapid publication in the scientific press (particularly the *Philosophical Magazine*). But generally, papers presented to the Society would, if substantial enough, eventually receive full publication in its lavishly illustrated *Transactions*; but the publication delays were so long that by the time they were published they had generally lost their novelty value in the debate, becoming little more than monuments to their authors. As in present-day science, such finished products of research were in any case highly stylized and artfully 'objective'. Yet when read in the context of the debates to which they were designed to contribute, it is not difficult for the historian to read 'between the lines', as knowledgeable contemporaries certainly did, and to discern there the continuing trace of the process of argumentative persuasion.

Thus throughout the continuum, from the most private to the most public forms of documentation, the historian has ample resources from which to reconstruct the intrinsically social processes by which a new piece of claimed knowledge, in this case the 'Devonian system' of strata and 'Devonian period' of earth history, came to be constructed and consensually validated.

A 'Coarse-grained' Analysis

The Devonian controversy arose in the course of development of 'normal science'. The dominant cognitive enterprise within geology in the 1830s was an attempt to order the sequence of strata in the Earth's crust. The increasingly explicit goal of this collective effort was to reconstruct the history of the Earth and of life on Earth before the advent of human beings.[35] A variety of techniques had been developed for this purpose, based above all on the careful detection of the correct sequence of the strata in particular regions, and the attempt to 'correlate' such sequences in different regions by matching their characteristic rock types and fossils. How in practice such techniques were to be employed was very much a matter of the shared tacit knowledge of geologists. Indeed, geologists' mutual evaluations of competence, and hence their ascribed places on the tacit social topography of the science, were largely determined by their degree of perceived success in applying this kind of unformalizable craftsmanship.

The collective application of these geological techniques had led to the compilation of a sequence of major groups or *'systems'* of strata.[36] Much of this sequence was accepted consensually as valid throughout the well-surveyed regions of Western Europe, and there were many who believed its validity might eventually extend worldwide. This collective enterprise had been less successful, however, in dealing with the lowest and therefore oldest strata. These typically rose to the surface in upland or mountain regions; they were usually crumpled, fractured or distorted; and they were generally lacking in easily distinguished rock types or well-preserved fossils. Some geologists feared that these oldest strata would remain forever a realm of 'chaos' — the epithet was their own. But to others such rocks presented a challenge to their skill and perseverance; for if they could produce order out of chaos *here* — if they could unravel the correct sequence and demonstrate its validity over a wide area — they would gain the credit for deciphering the earliest 'chapters' of the history of the Earth and of life itself.

In the early 1830s the London geologist Roderick Murchison was prominent among those pursuing this risky but valuable prize. In the Borderland between Wales and England he detected a group of these so-called 'Grauwacke' or 'Transition' strata lying in an unusually clear sequence below the well-known 'Carboniferous' system with its valuable coal deposits; and he found that the older strata contained distinctive animal fossils but no land-plant fossils at all. In the visual language of geology (then and now), his conclusions can be represented as a columnar 'section' or diagrammatic pile of the strata involved (Figure 2, column 1). He termed

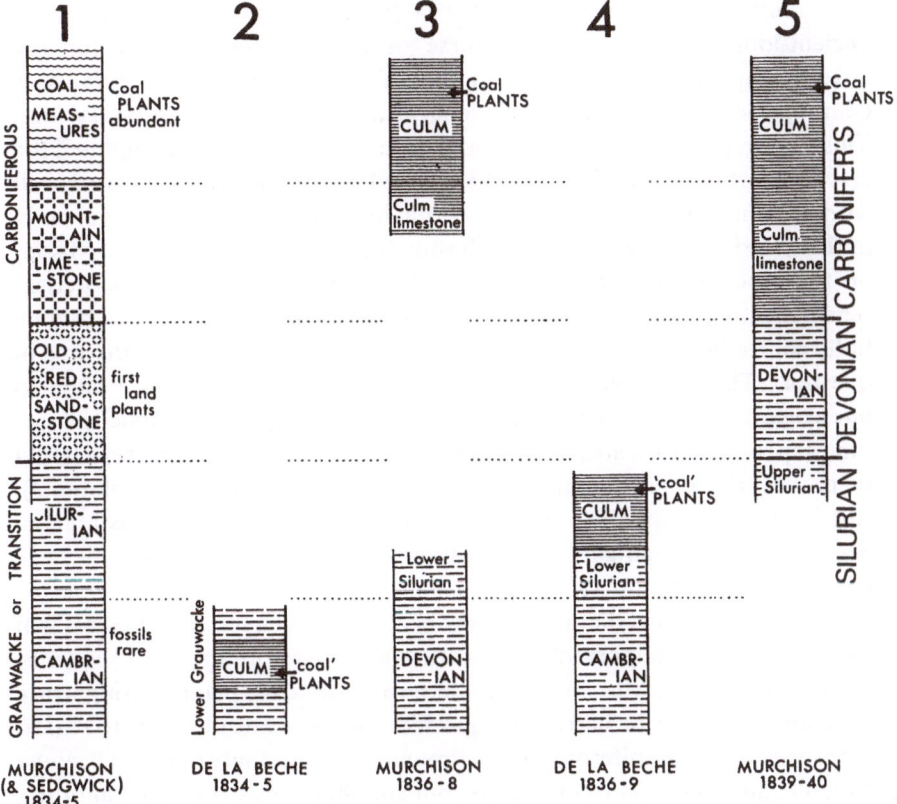

Figure 2. Diagrammatic representation of four successive interpretations of the sequence of strata in Devonshire (columns 2 to 5), matched or "correlated" with the "standard" succession in central England as extended downwards by Murchison (and Sedgwick) in Wales and the Welsh Borderland (column 1). The thicknesses of strata in these columnar ("vertical") sections are not to scale: the shadings represent various predominant types of rock.

'his' strata *Silurian*, after the tribe of Silures who had lived in the Borderland in Roman times.[37] His growing confidence in the general validity of the Silurian as a major period in the Earth's history was marred only by persistent reports from several other countries that similar ancient or pre-Carboniferous rocks contained abundant plant fossils and even coal seams. But in his project to establish the Silurian system, he did not feel inhibited by such apparent anomalies, until one was reported close to his own territory, and in a particularly serious form.

In southwest England, Henry De la Beche discovered plant fossils in ancient-looking strata in the course of his geological survey for the government, and these fossils were identified by a competent specialist as being species well known from the coal strata of the Carboniferous system. This report, presented late in 1834 at the Geological Society, precipitated the Devonian controversy. De la Beche insisted that the fossils came from well within an unbroken sequence of strata that had long been regarded as ancient in date, and certainly pre-Carboniferous (Figure 2, column 2). Murchison, on the other hand, felt so confident about his own emerging theoretical scheme, with the plantless Silurian system below the Carboniferous, that he flatly contradicted De la Beche, although he had never yet studied the area in question. The fossil plants, he claimed, must have come from a hitherto unrecognized patch of coal strata of Carboniferous age. This implied a huge time gap, and hence a structural discontinuity or 'unconformity', separating them from the genuinely ancient strata of Devonshire (Figure 2, column 3). But De la Beche insisted, and continued to insist, that there was no sign whatever of any such gap or discontinuity.

When eventually, in 1836, Murchison did go to Devonshire to look for himself, he concluded that De la Beche had indeed been seriously mistaken in the way he had deciphered the structure and therefore the local sequence there. According to Murchison's new interpretation, the plant-bearing strata were uppermost, as he had earlier guessed, so that it was plausible to regard them as truly Carboniferous. This eliminated the anomaly that had become so damaging to his Silurian scheme; but simultaneously it created another, for he had to gloss over his failure to find any trace of the gap or 'unconformity' that his interpretation required (Figure 2, column 3).

At the 1836 meeting of the British Association, Murchison gleefully announced De la Beche's gross 'mistake' in deciphering the structure of Devonshire. De la Beche was furious because, when made in such an inappropriately public arena, this claim was bound to lead the politicians to doubt his competence, and it thereby jeopardized his future employment. But in the cognitive dimension De la Beche soon accepted his critics' interpretation of the *sequence*, though not of its *dating*. Since he was convinced there was no perceptible gap in the sequence, and the plant-bearing strata were now to be placed at the top, he inferred that they might be Silurian (Figure 2, column 4). But of course this put them back just where they most seriously compromised Murchison's conception of his Silurian system.

Murchison wrestled unsuccessfully with this recalcitrant anomaly for two or three years more, until, in 1839, he suddenly announced a radically new interpretation (Figure 2, column 5). Tacitly accepting what De la Beche had maintained all along, namely that there was no gap in the sequence in Devonshire, Murchison proposed that the older strata were much less ancient than all the relevant geologists — himself included — had hitherto inferred. Instead, he now assigned most of them to a position (and age) equivalent to a distinctive group of strata — the 'Old Red Sandstone' — found at the base of the Carboniferous system in other parts of Britain. In the eyes of other geologists, however, this seemed no more than an ad hoc device to get Murchison off his own theoretical hook. For what he now termed 'Devonian' strata (incidentally changing the meaning of that term from an earlier usage: compare Figure 2, column 3) were radically different from the Old Red Sandstone in both rock types and fossils, and therefore made a highly implausible 'correlation'. Murchison now argued that it was the Old Red Sandstone strata, not the Devonian rocks, that were atypical of the period of their formation. The Devonian strata, he argued, *must* be of that intermediate age because the relevant specialists had pronounced their fossils to be intermediate in character between those of his Silurian system below and those of the rest of the Carboniferous system above.[38]

However, this implausible interpretation did embody an explicit set of predictions; and in the following two field seasons Murchison undertook extensive field work in Belgium and Germany, and then in Russia, to search for less ambiguous sequences of strata. He hoped these would establish the validity of what he came to regard as a full-fledged Devonian 'system', sandwiched between his Silurian system below and the now redefined and narrowed Carboniferous system above. Despite many setbacks, local anomalies and periods of doubt about this scheme, he eventually produced evidence that more and more of his geological colleagues found convincing. The 'Devonian system' of strata came to be consensually accepted as globally valid, and as representing a corresponding 'Devonian period' in the history of the Earth and of life. So after about a decade of vehément argument, the Devonian controversy subsided in the later 1840s, with only a handful of marginal men refusing to accept the new interpretation.[39] The Devonian system (and period) became a piece of increasingly solid scientific knowledge. Its solidity has been greatly enhanced in the subsequent century and a half by the consistent filling in of its detailed contents on a worldwide scale, in a way that would surely have impressed, but not surprised or perplexed, those who first proposed and supported the Devonian interpretation.

This brief summary of the Devonian controversy bears scarcely any trace of my earlier emphasis on the varied sources of evidence for analyzing the construction of the 'Devonian' as a piece of new knowledge, nor of my emphasis on that construction as a result of social processes spread across the diversified topography of geologists. It is a 'coarse-grained' summary that could have been written almost entirely from an analysis of formal published papers; and it describes the controversy almost entirely in terms of an argument between just two protagonists, Murchison and De la Beche.

Even in this crude form, however, the summary already shows one important feature which rarely appears in the idealized and formalized accounts of science that philosophers generally use for their own purposes. The interpretative scheme for Devonshire geology that was ultimately judged 'successful' in terms of its increasingly fruitful use worldwide, namely Murchison's proposal of a Devonian system (Figure 2, column 5), did not simply triumph over an earlier scheme with lesser explanatory power. It was a scheme that had not been foreseen by anyone, certainly not by Murchison himself, when the controversy started. Yet it was no simple compromise between De la Beche's scheme that precipitated the controversy (Figure 2, column 2) and the incompatible rival scheme that Murchison devised in response (Figure 2, column 3). De la Beche modified his first scheme to incorporate explicit concessions to Murchison's arguments (Figure 2, column 4), and Murchison modified *his* first scheme to embody similar — but only tacit — concessions to De la Beche's arguments (Figure 2, column 3 is a composite representation of Murchison's earlier and later schemes). The finally 'successful' scheme (Figure 2, column 5) was significantly different from *either* of the two earlier rival schemes, even in their modified forms. It tacitly absorbed De la Beche's continuing insistence that there was no evidence for any discontinuity in the sequence. But it also embodied Murchison's continuing insistence that fossil plants so similar to those of the coal deposits elsewhere must be of truly Carboniferous age, and not Silurian or even older. Over and above those elements inherited from the earlier rivals, however, the finally 'successful' scheme also introduced a radically novel element. This was the claim — initially a highly implausible one — that the older strata of Devonshire were much younger than all the geologists concerned had until then supposed. The perceived plausibility of this 'Devonian' scheme was strengthened only as the predictions it embodied were successfully fulfilled elsewhere in Europe.

In crude or 'coarse-grained' outline, therefore, the Devonian controversy illustrates a dynamic pattern of theory construction that may be much more common than either philosophers or historians of science have generally recognized. The 'successful' scheme emerged from a process of dialectical or mutual modification of incompatible rival schemes, yet it transcended them in a way that made it far more novel than any compromise, and in a way that was quite unexpected at the outset and highly implausible when first proposed.

Toward a 'Fine-grained' Analysis

The crude simplicity of my 'coarse-grained' summary of the Devonian controversy gives no hint of the astonishing complexity that emerges once the full range of documentary evidence is brought to bear on the reconstruction. The formal published papers of the chief participants then appear merely as the deceptively isolated tops of far larger icebergs. To change the metaphor to a more appropriate one, the published texts are the occasional press releases that emerge from a long and complex process of diplomatic negotiation behind closed doors. This process of negotiation can only be reconstructed by a detailed analysis of all its documentary traces: above all in the network of correspondence that passed between all the participants; but supplemented by the records of their field work, by reports of their arguments in the semi-public arena of the Geological Society and elsewhere, and — with appropriate reading 'between the lines' — by the texts of their formal papers. Only such a 'fine-grained' reconstruction will ever give an adequate picture of the dynamic processes of argument and persuasion that lay behind the construction of the Devonian system as a new piece of solid and reliable scientific knowledge.

Pending such a detailed narrative and its analysis, I can at least mention three examples of the ways in which the full documentation enlarges and transforms the coarse-grained summary. My examples are themselves of increasingly fine 'grain' or degree of temporal resolution. First, we can trace the formation of a consensus on the finally 'successful' interpretative scheme among six British members of the small core set of highly competent geologists. When the controversy broke out, they were divided on the issue into two unequal parties. De la Beche's scheme (Figure 2, column 2) was found plausible by three other influential elite geologists: Greenough, virtually the founder of the Geological Society; Buckland, who taught geology at Oxford; and Sedgwick, his counterpart at Cambridge. Murchison's objection to De la Beche was upheld by only one comparably important elite figure, namely Lyell, the self-appointed chief theoretician of British geology. The two parties were divided in effect by the confrontation of two apparently irreconcilable propositions. These can be summarized crudely in slogan form as, respectively, 'No gap in the sequence of Devonshire strata!' and 'No Carboniferous plants in any ancient strata!'

The subsequent course of negotiations, as revealed by the mass of precisely dated documents available, shows the importance of Sedgwick's gradual defection to Murchison's side, firmly established by their joint field work in Devonshire in 1836. It shows the complex modifications of the schemes proposed by some of these individuals, embodying certain concessions to their opponents in the light of empirical evidence that they had to admit as compelling. It also shows the quite rapid consensus of the core-set around the 'Devonian' interpretation, within about a year of its first public proposal by Murchison. But this process of negotiation and consensus formation, if conceived in terms of the cognitive trajectories of individuals, fails to give an adequate impression of the interactions between them.

It is also difficult to bear in mind that the 'Devonian' scheme was *not* a compromise between the initial rivals or even their later modifications, unless another aspect of the social process of negotiation is also emphasized. This is that a new 'conceptual space' opened up in the course of time between the positions that had earlier seemed irreconcilable. The two slogans in which I summarized those embattled positions were not of course logical opposites; but the possibility of a noncontradictory 'space' between them only became apparent through the intervention of individuals who were *not* members of the core-set of elite geologists, but men whose tacit status in the social topography of the science was only that of accredited geologists. More specifically, it was the young Devonshire geologist Robert Austen, and the fossil specialist William Lonsdale, the Geological Society's curator, who first produced evidence and opinions that implicitly opened up the new conceptual space that made the 'Devonian' scheme a conceivable possibility.

It was Murchison who eventually did conceive that possibility; and judging from the published record alone, it might appear to be a classic example of a sudden 'Eureka-experience' or *Aha-Erlebnis*. But — and this is my second example — a finer-grained reconstruction of the negotiations that led to his public proposal of the 'Devonian' scheme reveals a far more complex process of argument and persuasion. The cognitive pathway from Murchison's earlier scheme (Figure 2, column 3) to the new one two and a half years later (Figure 2, column 5) was far from straight. It involved a complex sequence of moves — proposals, counterproposals, new empirical reports, arguments about their significance, and so on — made by a varied cluster of individuals. These ranged from Murchison himself, with Sedgwick, his fellow-member of the elite core-set, acting as a kind of friendly but critical sparring partner, through accredited geologists like Austen and Lonsdale with local or specialized expertise, to merely amateur geologists like the Devonshire country gentleman William Harding, who discovered significant new fossil plants at a crucial moment in the intricate process of argumentative negotiation. If one follows Murchison's 'Devonian' scheme with strict attention to chronology and abstains from the dubious benefits of hindsight, it can be seen to have emerged in a quite unpredictable way from a sequence of incomplete and inconclusive arguments involving many others besides him. It is this social process of complex interactive negotiation that makes it legitimate to term the ultimately 'successful' Devonian interpretation a social *construction*.

Turning a still higher power of historical magnification onto this process of negotiation, for my third and last example, almost any substantive letter between members of the core-set in the Devonian controversy would demonstrate the argumentative, persuasive and rhetorical character of their interactions. A particularly vivid example is one where the message was conveyed in visual as well as verbal form. When De la Beche sent the Geological Society specimens of the fossil plants he had found in strata of ancient ('Grauwacke') appearance in Devonshire, the reliability of his report was immediately queried by Murchison, backed up by Lyell, although these critics had not seen for themselves the rocks and the locality De la Beche had described. As already mentioned, it was this incident that precipitated the Devonian controversy. De la Beche heard reports of the argument after the meeting, and was understandably incensed that his competence had been impugned in his absence. He wrote at once to both Greenough and Sedgwick, enclosing a caricatured version of what had occurred (Figure 3). He depicted himself standing before the other members of the Geological Society, and declaring "This, Gentlemen, is my *Nose.*" They, however, answered him, "My dear fellow! — your account of yourself generally may be very well, but as we have classed you, *before we saw you*, among men *without noses*, you *cannot possibly have a nose.*"[40]

This superficially innocent joke was pervaded with serious rhetorical purpose: it was a carefully devised move in the larger field of persuasive negotiation. De la Beche entitled his caricature "Preconceived Opinions *v[ersus]* Facts," in order to appeal to the well-known 'Baconian' ideals of his two correspondents. By contrast, he depicted his critics viewing the evidence of their senses through the colored spectacles of illegitimate theorizing. Furthermore, he showed himself dressed in the rough topcoat of a field geologist, ready to brave all weathers to study the raw facts of nature, whereas his critics were dressed in elegant tailcoats more suitable for purely speculative debate in a fashionable London *salon*. He sent this iconographically loaded caricature to Greenough, already his patron and champion in London, knowing full well that Greenough would show it

Figure 3.   Caricature by De la Beche, criticizing Murchison's and Lyell's reaction to his report
(see Figure 2, column 2) of fossil plants of Coal Measures species in the 'Grauwacke' strata of
Devonshire.
Caption: Preconceived opinions vs. facts
De la Beche: This, Gentlemen, is my *Nose*.
His critics: My dear fellow!— your account of yourself generally may be very well, but as we
have classed you, *before we saw you*, among men *without noses*, you *cannot possibly have a nose*.

round among Geological Society members, and hoping that it would
strengthen collective opinion in his favor before the controversy went any
further, as he knew it was bound to do. He sent a separate copy to Sedgwick
in Cambridge, hoping to undermine the working alliance that had grown up
in other areas of research between Sedgwick and Murchison, and knowing
that Sedgwick would also read the message as supporting implicitly his own
criticism of Murchison's other ally, the theoretically overambitious Lyell. In
this way, read in context, De la Beche's joke can be seen to have had the
serious rhetorical purpose of altering the balance of forces in the incipient
controversy in his own favor, particularly by isolating his critics Murchison
and Lyell from their most influential potential ally, Sedgwick. But what De la
Beche exceptionally put into visual form is no different from what he and all
other geologists involved in the controversy put into verbal forms whenever
they wrote to each other, and no doubt also in their face-to-face
conversations of which the historian has no direct trace.

The entire web of their interactions constitutes what Latour and Woolgar,
in their 'anthropological' study of *Laboratory Life* in modern

neuroendocrinology, have termed the 'agonistic field' of persuasive forces: forces which continually push in different directions the perceived plausibility of rival theoretical suggestions.[41] The main schemes of interpretation that I outlined in my coarse-grained summary of the Devonian controversy acted like locally 'gelled' portions of the agonistic field; they were relatively solid and secure pieces of plausible knowledge, candidates for further solidification, yet with enough defects for their dissolution into implausibility to be a conceivable outcome too. But when the 'Devonian' scheme gained the consensual support of all the elite geologists involved, with only a few marginal men holding out against it, this did *not* constitute a unique and irreversible 'inversion' into a spuriously solid scientific 'fact', as Latour and Woolgar suggest for their own case study.[42] The Devonian system and period became a more solid or 'fact'-like piece of knowledge only through proving itself valid *in use* over a progressively wider range of regions and in progressively greater and more consistent detail. It was entirely conceivable at the time that it might have failed in this test of use; it might have declined in plausibility and fallen into oblivion, as other comparable schemes had done, by proving inapplicable or inconsistent beyond the regions where it was first constructed.

That the Devonian system and period did prove so successful in descriptive and explanatory use must surely raise grave doubts about the extreme epistemological scepticism which is adopted by many of those currently working in the cognitive sociology of science.[43] The Devonian controversy is a particularly cogent example in this debate, because one can readily agree that in this case the entity produced by the social processes I have described was itself an *artificial* or human construct.

The Devonian system and period were not simply 'discovered' as natural objects. The total sequence of strata in the Earth's crust and the continuum of the Earth's entire history could in principle have been divided for descriptive purposes in many other, equally useful and valid ways. It is clearly a matter of historical contingency — a complex result of the European location of the relevant pioneer research — that the strata (and the history they represent) have come to be divided in the way they are, with the Devonian system (and period) defined and delineated as a particular portion. Nonetheless, that portion has been (and is) claimed to represent a part of the unique and real history of the Earth, however imperfectly described and understood.

I conclude that a historical example such as the Devonian controversy, if reconstructed and analyzed in fine-grained detail, fully confirms the view that scientific knowledge is *constructed* through a complex *social* process of interactive negotiation, among individuals of differing competences holding widely divergent initial opinions. But within this process of social construction, the common currency in the negotiations is that of the empirical phenomena that are consensually accepted as relevant and ultimately binding, however ambiguous their interpretation may initially seem to be. In other words, the natural phenomena constitute an input that has a constraining and differentiating effect on the outcome of the social processes; and the entity that is socially constructed is thereby a mediated representation of an external reality that exists independently of those who construct it. But if philosophical realism is to be revived in the analysis of scientific knowledge, as I believe it should be, then it must be in a form that acknowledges that the human learning processes that generate reliable natural knowledge are *social* in character through and through. The individual scientist confronts the natural world only through the mediation of group interactions in the social world.

Notes

Earlier presentations of similar material were given as a Special University Lecture in London in January 1981 and at the ERISS conference ("Epistemologically Relevant Internalist Sociology of Science") organized by Syracuse University in June 1981. For valuable comments and criticism on these and other occasions I am especially indebted to David Bloor, Donald T. Campbell, Gerald L. Geison, Patrick de Maré, J.B. Morrell and Sylvan S. Schweber.

1 Peter Medawar, *The Art of the Soluble* (London: Methuen, 1967), p. 151.
2 An important and characteristic example is Bruno Latour and Steve Woolgar, *Laboratory Life: The Social Construction of Scientific Facts* (Beverly Hills and London: Sage, 1979). Latour, who had earlier done anthropological field work in the Ivory Coast, was a participant observer at the Salk Institute, La Jolla, California, in 1975–77.
3 For example, Latour and Woolgar (*op. cit.,* note 2, pp. 106–7) disclaimed any intention of writing a *history* of the "Construction of a Fact" of neuroendocrinology; yet a history is precisely what they were obliged to write, in order to make sense of developments stretching back years before the period of participant observation.
4 For example, H.M. Collins's studies of claims to the observation of high-flux gravitational radiation: "The Seven Sexes: A Study in the Sociology of a Phenomenon, or the Replication of Experiments in Physics," *Sociology* 9 (1975) 205–224; and its sequel, "Son of Seven Sexes: The Social Destruction of a Physical Phenomenon," *Soc. Stud. Sci.* 11 (1981): 33–62. See also Trevor Pinch's study of the continuing debate on the observation of solar neutrinos: "Theoreticians and the Production of Experimental Anomaly: The Case of Solar Neutrinos," in: *The Social Process of Scientific Investigation*, ed. Karin D. Knorr et al. (Dordrecht, Boston and London: D. Reidel, 1981), pp. 77–106; and "The Sun-Set: the Presentation of Certainty in Scientific Life," *Soc. Stud. Sci.* 11 (1981): 131–158.
5 For example, David Kohn, "Theories to Work By: Rejected Theories, Reproduction, and Darwin's Path to Natural Selection," *Stud. Hist. Biol.* 4 (1980): 67–170; Sandra Herbert,

"The Place of Man in the Development of Darwin's Theory of Transmutation," *J. Hist. Biol.* 7 (1974): 217–258, and 10 (1977): 155–227; Silvan S. Schweber, "The Origin of the *Origin* Revisited," *ibid.* 10 (1977): 229–316.

6 This has been most explicit in studies by Howard E. Gruber: see for example his "The Evolving Systems Approach to Creative Scientific Work: Charles Darwin's Early Thought," in: *Scientific Discovery: Case Studies*, ed. Thomas Nickles (Dordrecht and Boston: D. Reidel, 1980), pp. 113–130; and *Darwin on Man. A Psychological Study of Scientific Creativity*, second ed. (Chicago: Chicago University Press, 1981), esp. chap. 8.

7 This claim must be distinguished from the assumptions that underlie most research using citation analysis: see David O. Edge, "Quantitative Measures of Communication in Science: a Critic Review," *Hist. Sci.* 17 (1979): 102–134.

8 In this article the focus is on the implications of the case for the historical sociology of scientific knowledge. It is designed to complement one that I wrote for an audience of geologists, which is more "technical" in content: M.J.S. Rudwick, "The Devonian: A System Born from Conflict," in: *The Devonian System*, ed. M.R. House et al. [London: Palaeontological Association (*Spec. Pap. Palaeont.* 23), 1979], pp. 9–21.

9 The phrase is contemporary, but is aptly used in J.B. Morrell and Arnold Thackray, *Gentlemen of Science. The Early Years of the British Association for the Advancement of Science* (Oxford: Oxford University Press, 1981).

10 "Gentlemen-specialists" is my own phrase: see Martin J.S. Rudwick, "Charles Darwin in London: The Integration of Public and Private Science," *Isis* 73 (1982): 186–206. See also Roy Porter, "Gentlemen and Geology: the Emergence of a Scientific Career, 1660–1920," *Hist. Jl.* 21 (1978): 809–836.

11 There is no adequate general history of geology in this period. For the immediately preceding period, see, however, Roy Porter, *The Making of Geology. Earth Science in Britain, 1660–1815* (Cambridge: Cambridge University Press, 1977).

12 George Bellas Greenough (1778–1855), cofounder and first president of the Geological Society of London, the first specialist society in the world to be devoted to geology.

13 The following passage is summarized from Rudwick, "Darwin in London" (note 10).

14 Only the socially marginal "scriptural geologists" — the distant intellectual ancestors of modern creationists — disturbed this general consensus, in that they claimed that they, and not the mainstream geologists, were the true fount of authentic knowledge about the Earth and its history. Leaving them aside, however, the otherwise consensual form of this graded topography can be described quite simply. The scriptural geologists still await their historian: for a preliminary survey, see Milton Millhauser, "The Scriptural Geologists. An Episode in the History of Opinion," *Osiris* 11 (1954): 65–86.

15 Individuals could of course move quite rapidly across this topography — Darwin's early career is a good example: see Rudwick, "Darwin in London" (note 10), Figure 1. Furthermore, on a longer time-scale the form of the topography itself changed gradually, for example with the increasing specialization and esoteric character of the subject matter of the science. Any such topography is therefore an attempt to represent in visual form a set of contingent circumstances that were specific to a particular branch of science at a particular period.

16 "Focal problem" is my own term: see Rudwick, "Darwin in London" (note 10). The "hot spot" metaphor is used by H.M. Collins, "The 'Core-set' in Modern Science: Social Contingency with Methodological Propriety in Science," *Hist. Sci.* 19 (1981): 6–19.

17 See Collins, "Core-set" (note 16).

18 For example, Darwin was undoubtedly in the core-set that was simultaneously debating the reality or otherwise of the widespread elevation of continental land masses within recorded human history; but he played only a marginal role in the Devonian controversy. See Martin J.S. Rudwick, "Darwin and the World of Geology," in: *The Darwinian Heritage*, ed. David

Kohn (Chicago: Chicago University Press, forthcoming) and "Darwin in London" (note 10).

19 "Information flow' is used here in an information-theory sense; it does *not* imply that what flowed were theory-free empirical observations.

20 On the two societies, see Rudwick, "Darwin in London" (note 10), pp. 192–3. The scriptural geologists were almost equally strong-boundaried, although not organized formally; they had a mutually hostile frontier against — particularly — the Geological Society of London.

21 The terms "arena" and "social drama" are particularly apposite to convey the dramaturgical and (even) gladiatorial character of some of these discussions: see Rudwick, "Darwin in London" (note 10), and James A. Secord, *Cambria/Siluria: the Anatomy of a Victorian Geological Debate*, Princeton University Ph.D. dissertation, 1981.

22 See Morrell and Thackray, *Gentlemen of Science* (note 9), chap. 3.

23 *Ibid.*, pp. 462–5; also Rudwick, "Devonian" (note 8).

24 This point is rightly emphasized in — and is indeed central to — recent "anthropological" research on modern scientific practice (see note 2); but I do not follow such writers in their radical epistemological scepticism (see below).

25 For an analysis of a modern example, attributing the systematic remodeling to the needs of public presentation, see Steve Woolgar, "Discovery: Logic and Sequence in a Scientific Text," in: *Scientific Investigation* (note 4), pp. 239–268. A famous historical example is Darwin's recollection, late in life, that he had worked on "Baconian principles" and "without any theory" [*The Autobiography of Charles Darwin, 1809–1882*, ed. Nora Barlow (London: Collins, 1958), p. 119], when compared with the evidence of his 'species notebooks' that he was theorizing in a wide-ranging — and of course highly creative — way at the relevant period (see references in notes 5, 6).

26 The following passage amplifies my brief description of such a continuum, in Rudwick, "Darwin in London" (note 10).

27 For the importance of analyzing laboratory notebooks, see F.L. Holmes, "The Fine Structure of Scientific Creativity," *Hist. Sci.* 19 (1981): 60–70. His forthcoming major study of the work of Hans Krebs is represented in preliminary form in "Hans Krebs and the Discovery of the Ornithine Cycle," *Proc. Fed. Amer. Soc. Exper. Biol.* 39 (1980): 216–225. As far as I know, there are as yet no comparably detailed studies based on geological field notebooks.

28 Victor Turner extends and generalizes van Lennep's classic concept of the liminal to include a wide range of 'liminoid' situations: see *Dramas, Fields and Metaphors. Symbolic Action in Human Society* (Ithaca and London: Cornell University Press, 1974). One of his major examples, that of pilgrimage, illuminates in my opinion the phenomena of geological field expeditions.

29 Some of the most tantalizing gaps in the extant correspondence are due to the disappearance of certain series of letters *since* the compilation of the relevant late Victorian "Lives and Letters," which only printed incomplete and unreliable extracts.

30 For example, of the major actors in the Devonian controversy, Adam Sedgwick (1785–1873) was tied for much of the time by academic and ecclesiastical duties to Cambridge and Norwich; Henry De la Beche (1796–1855), by his governmental surveying duties, to Cornwall and South Wales; William Buckland (1781–1856), by academic duties, to Oxford; whereas Roderick Murchison (1792–1871), Charles Lyell (1797–1875) and Greenough (note 12) were based in London. They all met fairly frequently in London, however, particularly at meetings of the Geological Society and over the convivial dinners of its Club.

31 For example, Sedgwick and Murchison sometimes exchanged letters every day of the week by the overnight mail between London and Cambridge (or Norwich). Letters between London and the furthest end of Cornwall were delivered in two days, and those between London and Paris or Bonn took little longer. The high cost of postage (until the introduction

of the penny post in 1840, while the Devonian controversy was still raging) did not greatly deter these gentlemen of science.

32 Some sociologists of science have recently argued that, because all statements contained in such documents are necessarily indexical, i.e., dependent on their context for their meaning, no interpretative reading is ultimately more reliable than any other. My argument is that this conclusion does not follow from its premise, and that specific readings can become more reliable than others, the more fully their original context is reconstituted.

33 I use "trans-personal" in the sense in which it was originally developed to describe the dynamics of human groups in therapeutic contexts: see for example P.B. de Maré, *Perspectives in Group Psychotherapy* (London: Allen and Unwin, 1972). Some such term is undoubtedly needed to do justice to the historical reports of a "sense of consensus" on certain occasions.

34 See Morrell and Thackray, *Gentlemen of Science* (note 9), pp. 139–148. Several major actors in the Devonian controversy used these annual meetings regularly to rehearse arguments they planned to present later to a more expert audience at the Geological Society.

35 The originally *structural* and therefore *non*-historical goal of what is now termed stratigraphy is reflected in its conventional forms of visual representation: see Martin J.S. Rudwick, "The Emergence of a Visual Language for Geology, 1760–1840," *Hist. Sci.* 14 (1976): 149–195. This structural style is characterized in Rudwick, "Cognitive Styles in Geology," in: *Essays in the Sociology of Perception,* ed. Mary Douglas (London: Routledge and Kegan Paul, 1982), pp. 219–241.

36 The modern sense of the term "System," as a major assemblage of strata representing a specific period of time and therefore recognizable (in principle) anywhere in the world, came into use partly as a consequence of the outcome of the Devonian controversy. During and before that time, the term was used in many different senses, each with its own theory-loading.

37 His colleague Adam Sedgwick, professor of geology at Cambridge, termed still older but much more obscure strata "Cambrian," after the Roman name for Wales. Their famous long quarrel over their respective "Systems" was only indirectly related to the Devonian controversy; it is described in detail in Secord, *Cambria/Siluria* (note 21).

38 The fossils of these two Systems only became well known, and therefore available for this kind of comparison, *during* the Devonian controversy itself, with the publication of John Phillips's *Illustrations of the Geology of Yorkshire, Part II* (1836) and Murchison's own *Silurian System* (1839). The latter relied on fossil identifications by specialists such as William Lonsdale (1794–1871) and James de Carle Sowerby (1787–1871), who, with Phillips, were also responsible for evaluating the fossils from the disputed Devonshire strata.

39 The most vocal of the residual sceptics was the Somerset clergyman David Williams (1792–1850), who had contributed actively to the controversy on the basis of detailed, but only local, geological knowledge of southwest England. De la Beche and Greenough conceded the validity of the "Devonian" interpretation in substance, while saving face by avoiding using the term "Devonian System" itself.

40 De la Beche's scientific caricatures, some of them lithographed by himself to give them wider circulation, were well known at the time. Several are reproduced in Paul J. McCartney, *Henry De la Beche: Observations on an Observer* (Cardiff: Friends of the National Museum of Wales, 1977). The successive drafts that led to one famous example are analyzed in Martin J.S. Rudwick, "Caricature as a Source for the History of Science: Henry De la Beche's Anti-Lyellian Sketches of 1831," *Isis* 66 (1975): 534–560.

41 Latour and Woolgar, *Laboratory Life* (note 2), chap. 6.

42 *Ibid.*, chap. 4.

43 *Ibid.*; see also most of the essays in *Scientific Investigation* (note 4) and *Knowledge and Controversy, Studies of Modern Natural Science*, ed. H.M. Collins, special issue of *Soc. Stud. Sci.* 11 (1981): 1–158.

On the Devonian Controversy
A Comment

SILVAN S. SCHWEBER

Professor Rudwick in his paper has given us a display of what history of science in like at its best: a mastery of the technical aspects of the subject, a sensitivity for the complexity of historical events, and all this conveyed in a language that reflects its author's appreciation of the aesthetics of style. Furthermore, I believe he has charted useful directions for conveying the richness of the materials he has dealt with. As in his past writings, his emphasis on the visual and visualization has helped us see better the topography of the social and sociological scene that he is mapping. I find the charts that he has devised insightful and very helpful; and I believe it would be very worthwhile to explore further means of representing visually the parameters that are being invoked to describe historical change. And I have not praised Professor Rudwick in order to be critical in my further remarks. What has been presented to us is an important and seminal approach to the history of science. But Professor Rudwick has attempted to do more than write elegant, very good history of science understood in the widest sense of intellectual history. He has also used his historical research into the Devonian controversy to argue a case. The case is that scientific knowledge is constructed through a complex social process of interactive negotiations, among individuals of different competences holding widely divergent initial opinions; that the natural phenomena under discussion constitute an input

E. Ullmann-Margalit (ed.), The Kaleidoscope of Science, 219–223.

that constrains and narrows the options on the outcome of the social processes; and that "the entity that is socially constructed is thereby a mediated representation of an external reality that exists independently of those who construct it."

No one could argue with Professor Rudwick's first point that science is a social process, though I find that the word 'social' is being used with variously differing meanings. In one sense — the sense most often used by Professor Rudwick — it denotes the interactions between the various actors engaged in the drama that constitutes the production of 'scientific' knowledge. It is a special kind of drama, because although the subject matter and the stage props ('the natural objects') are prescribed, the actors, to a certain extent, can write their own play and chose their own parts, provided they abide by certain rules. The aim is to devise a script to which all the actors can agree and an arrangement of the props and sets that is mutually acceptable.

There is also a second sense in which Professor Rudwick has used the word 'social' in his description of "science as a social construction" and that is as synonymous with 'external'. Though only vaguely referred to in this paper, in his writings on the formation of the Geological Society[1] and elsewhere,[2] Professor Rudwick in fact has given a clear indication where these 'external' social influences manifested themselves in the delineation of what constituted acceptable scientific theories to account for geological phenomena, and how these 'social' constraints placed certain boundaries on the debates which could be held. The other linkage through which the 'social' enters the picture, namely the psychological, in terms of the ambitions, interests, and character of the various actors, is also implicit in Professor Rudwick's presentation. In all these points I am in full agreement with him. The issues I want to address are concerned with the aim of the exercise, which I take to be more than describing an episode in the unraveling of the history of the Earth. The object of the enterprise is to learn something about the growth of knowledge in the natural sciences, to learn more about the construction of "the external reality that exists independently of those who construct it," and to be able to discern and characterize the "constraining and differentiating effect" of natural phenomena.

It is on this score that I have some difficulties with Professor Rudwick's paper. For I believe that, if we are to learn more about the characterization and the growth of knowledge in the sciences and its linkage to the social context, we shall have to differentiate between the various sciences much more carefully. It is not merely that different sciences deal with different entities, but also that the nature of the objects they deal with gives rise both to different kinds of theoretical descriptions and to different criteria of what constitutes adequate theories.

We should also distinguish between the different stages in the evolution of the various sciences. Every science deals with a certain fairly well-delineated universe made up of certain entities, its ontology. Once a science attains the stage at which the theoretical structures match the ontological structures, a certain important stage in its development has been reached. The stability of these epistemological and ontological structures — as well as the stability of the fit — determine the importance of the stage. I will use that much abused word 'paradigm' to denote the cognitive structures that a community possesses once such a match exists — no matter how delicate or unstable the stability of the match, or of the structures themselves.

From such a perspective the Devonian controversy and the ensuing clarification, though surely "born from conflict" as Professor Rudwick has convinced us, is, I believe, a fairly typical manifestation of the activities going on in 'normal science'. Even though at the end of the controversy a new stratum, the Devonian, was inserted and a new chronology established, the conceptual framework of geology was nevertheless not altered. Moreover, the very way the conflict was resolved — using standard accepted procedures of correlating fossil content — gives further credence to this view. Professor Rudwick has given us a revealing keyhole look into the dynamics of adjudicating the conflicts and tensions, the role played by the core-set and the various other participants in a fairly typical episode of normal science.

I am of course open to the suggestion that we are seeing more than that: that the introduction of the Devonian constituted a fundamental, discontinuous restructuring of the ontology of geology, or of its theoretical structures, but that case, I believe, has not been made. It would seem to me that the very stability of the membership of the core-set during the entire episode suggests that indeed we are only seeing normal science. It would in fact be interesting *to correlate membership in core-sets* and passage to and from elite status, *with the importance of the case study* being depicted and analyzed as far as the growth of knowledge in the particular discipline is concerned (i.e. revolution, mini-revolution, readjustment of 'paradigm', etc. ...).

What underlies my criticisms — and they are addressed more against Kuhn and some of the Kuhnian presuppositions that Professor Rudwick has adopted — is the notion that what has been described is not only how things were, but how they should be, and how they are likely to be, together with the tacit assumption that all sciences are the same.

I would like to suggest that such case studies are valuable precisely because they will help us to understand how one scientific activity differs (and/or differed) from another; and by clarifying the way the conceptual framework of a science is reflected in 'social' activities, its modes of resolution of conflict, etc., to understand better the cognitive and other differences between different disciplines in any given period (e.g. chemistry and comparative anatomy).

The kinds of questions such case studies will have to address are: How did the nature of the experimentation (or exploration) affect time scales, costs, etc. and these in turn the social relations? How did experiments end? That is, what were the criteria which helped an individual decide that his experiment was 'completed', that what he had done or the information gathered could make a case? What were the criteria (intellectual and others) which decided that a temporary resolution had been achieved within a discipline, and what is similar and different in the various disciplines? All this is necessary to get a handle on the *progressive* development of the various sciences, on what constitutes progress, etc., and thus to gain insights into the nature of the "external reality that exists independently of those who construct it."

Let me conclude by making my point as explicit as I can. In the two sciences that I know a little about — physics and biology — something fundamentally new has happened this century, and I am referring to the formulation of quantum mechanics to account for atomic phenomena and to the elucidation of the role of DNA, RNA, and the genetic code in living organisms. These have fundamentally altered the character of these sciences. It is not that they have become *finalized* but rather that we have perceived and described certain *stable* levels in the hierarchical structures that seem to make up the world.

If I am right, this does imply a progression and that there are landmarks; that we are not merely historians but historians of *science*. It is not that I am advocating Whiggish history — far from it — but if this progress is real — and I believe it is — then to understand how these islands of stability came to be takes on selective importance. The task of the historian of science, as I conceive it, is to obtain both a diachronic and a synchronic view of the sciences, to understand their 'stability', their diversity and the dynamics of this diversity.

Notes

1   M.J.S. Rudwick, "The Foundation of the Geological Society of London: Its Scheme for Co-operative Research and its Struggle for Independence," *British J. Hist. Science* 1 (1963): 325–355.

2 M.J.S. Rudwick, "The History of Scientific Knowledge as a Social Construction: Implications for Theistic Belief," in: *The Sciences and Theology in the Twentieth Century*, ed. M.B. Hesse and A.R. Peacocke (London, 1982); see also his "Individuals and their Interactions in Science Past and Present: Introduction," *Hist. Sci.* 19 (1981): 2–5.

Knowledge and Power in the Sciences

EVERETT MENDELSOHN

In the fall of 1947, J. Robert Oppenheimer, the man who directed the United States project that made the atomic bomb, delivered the Arthur D. Little Memorial Lecture at the Massachusetts Institute of Technology. In the course of that lecture, he said:

> Despite the vision and far-seeing wisdom of our wartime heads of state, the physicists felt a peculiarly intimate responsibility for suggesting, for supporting, and in the end in large measure for achieving the realization of atomic weapons. Nor can we forget that these weapons, as they were in fact used, dramatized so mercilessly the inhumanity and evil of modern war. In some sort of crude sense which no vulgarity, no humor, no overstatement can quite extinguish, the physicists have known sin, and this is a knowledge which they cannot lose.

Physicists have known sin. Sin came to them, in Oppenheimer's view, through power, the power that science had gained and used.

Going back over 300 years to the beginning of the period which we refer to as the Scientific Revolution, we find that the image of power was put very clearly and very directly by Francis Bacon, philosopher, Lord Chancellor, a commentator on knowledge and what it might do. Let me quote him.

> It is well to observe the force and effect and consequences of discoveries. These are to be seen nowhere more conspicuously than in those three which were

225

E. Ullmann-Margalit (ed.), The Kaleidoscope of Science, 225–240.
© *1986 by D. Reidel Publsihing Company.*

unknown to the ancients and of which the origin, though recent, is obscure, namely, printing, gun powder, and the magnet. For these three have changed the world: the first in literature, the second in warfare, the third in navigation, whence have followed innumerable changes. In so much that no empire, no state, no star seems to have exerted greater power and influence in human affairs than these mechanical inventions.

"Knowledge is power," Francis Bacon said in his famous epigram. "We understand nature in order to command her" is another well-worn Baconian phrase. The goal was human gain, to improve the commonweal, but the basic aim was to have dominion over things. The new science, the new knowledge was, indeed, to improve the estate of humans on earth. But there was something more behind Bacon's science, and, in a sense, it had a power over him, as it was to have over his contemporaries. In developing this new approach to nature, these new ways of controlling nature, humans, he felt, would regain their prelapsarian state, the period before the fall. They could prolong life, they could conquer disease, they could do all those things which humans had dreamed of, but which they had not been able heretofore to achieve.

More than 300 years separate Bacon and Oppenheimer. The earlier vision of Bacon was an optimistic one, but curiously enough, it was developed in a time when humans actually possessed very little power to control nature. The more recent vision is a pessimistic one. At the very time that humans have achieved the ability to exert power over nature, they question the ends of such ability. What has happened in between? Why has this change of outlook and change of attitude occurred?

There are two clear senses in which the notion of power is used. One reflects science's ability to do powerful things, to bend nature to the human will. The second refers to the achievements of scientists and to their positions of power and authority in the societies of which they are a part. From where does science derive its power? How has it developed?

There are at at least four ways of looking at science and its sources of power. First, science is a body of concepts and techniques involving theories and ways of acting and thinking. Second, science is a way of knowing, a way of ordering reality and ultimately acting upon that reality. It provides a way of knowing that has transcended the boundaries of the practicing scientist and has·been adopted broadly in society at large. Third, science is a socially organized activity, a profession with a locus within a society and part of a hierarchy, related to other social institutions including governmental, military, and industrial. Finally, science is a source of utilitarian instruments of power for scientists and for others to use.

The seventeenth century was a period of fertile, conceptual innovation. Indeed, the innovations in ways of knowing and in conceptual structures of this period are what we label as the Scientific Revolution. An epistemology was established, which determined a way of knowing nature through reason and experience. Processes of demarcation were established to set rules of what was to be included in the sciences and what was to be excluded from them. The cognitive nature and the content of science were being developed.

Rules of exclusion from science were important. There were certain topics science would eschew. Science would not deal, said the founders of the Royal Society of London (and, indeed, it could have been repeated by the founders of almost any of the academies of science in the seventeenth century) with religion, rhetoric, metaphysics, politics, or morality. But even though it claimed it would not deal with these topics, science did maintain one value quite clearly — dominion over nature or mastery over things. While moral or normative guides were eschewed, the notion of mastery was central. Historians have called this the positivist compromise.

The seventeenth century was also the period when the social differentiation of the scientists' role was begun. Science was institutionalized and its organizations established. It became separated from other activities within society. Scientists could be recognized as a group called natural philosphers or experimental philosophers, but clearly separate from theologians, princes, and merchants. The institutionalization also provided a putative means of resolving conflict and of establishing truth. New scientific societies and academies, as they were established in the seventeenth century, provided the institutional bases within which new methods were developed. In the context of these societies debates about nature were held. The role of scientific peers was developed as challengers of findings, and the need to provide reproducible results was recognized. These societies were responsible for the introduction of processes of control — control which was both social and cognitive — over the ways scientists acted and the ways scientists thought. They established the content, boundaries, and procedures of science.

Individuals surrendered their authority to these societies at some level, while the societies, in turn, developed interests of their own. These institutional interests became reflected in the way scientists acted and in the way they responded to other interests of the society.

This was the period of the formation of ties with the state for support and for financial patronage. In Italy, France, Germany, and Russia, the state — or its agents, the aristocracy — were the direct patrons and controllers of science during much of the seventeenth and eighteenth centuries. Only in

England was the situation different. There, while scientists gained a Royal Charter in their early years, they gained nothing in the way of royal patronage or actual support. But as we look at the period of the seventeenth century, we can claim that this was a time when science became officially as well as informally recognized and empowered.

But what about power — Bacon's goal? During the seventeenth century, science's power either to command nature or to gain power of a social sort was extremely limited. Few were in science's debt, though science was in debt to many others. The interests of science, at this time, were largely those of an esoteric or organizational kind. Society had some specific interests in what science could do, say for example, in astronomy and its aid to navigation. But, by and large, the social interests surrounding science were the narrower ones of the group of people organized together to carry out the activity of science, including their social status. They were largely from the educated upper-middle classes, and they sought to achieve recognition within their society through their work. Science had proclaimed its utility to society, and its practitioners attempted to demonstrate this utility in order to enhance their roles. Furthermore, science, at the time, generally avoided conflict with the established power of state and church. Bishop Thomas Spratt's *History of the Royal Society*, written a few years after its founding in the 1660s, is largely a treatise of apologetics supporting the premise that science would in no way tread on the toes of the state or the church. We can conclude, therefore, that the cognitive features of science yielded little actual power. No sovereigns feared it, and several were involved in taming it. The church, of course, subdued it at its will, as the Counter-Reformation made abundantly clear.

The eighteenth century witnessed a further transformation of the scientific way of knowing, a transformation particularly involving the public. This change was largely outside the boundaries of science itself. Both the new politics of the Age of Reason and the new politics of the incipient Industrial Revolution gave legitimacy to the scientific way of knowing and to the belief that one could act on nature in an effective and economically viable manner. Scientists themselves were only partially cognizant of the new way of looking at nature that was taking shape outside their ranks as part of the development of the industrial societies. The sources of industrially oriented experimentation were examined by Henry Guerlac in his studies of the development of industrial chemistry in France during the eighteenth century. The people involved were at one and the same time practicing chemists and developers of the theoretical bases of what emerged by the end of the century as the Chemical Revolution. This same approach appears in the works of

Condorcet, the great philosopher of the French Enlightenment, who carried philosophical and educational visions into the revolution itself. Science represented, for him, the highest point of progress of the human race.

The provincial manufacturers in England also reponded to these new ways of dealing with nature. Textiles were their focus, and dyes and mordants were the chemical agents they needed. For leather-work and for metals they needed acids and alkalies. As a means of furthering interests in studying those sciences allied to new industries, they set up new organizations in the provincial centers — Manchester, Birmingham, Newcastle, Leeds. Through these new societies not only did they gain their practical view of what the sciences had to offer, but, as Arnold Thackray has shown regarding the Manchester Literary and Philosophical Society, they also began using science as the basis of a new industrial middle-class culture. They were, after all, outside the bounds of the upper classes of British society and, as such, divorced from the culture of the belles lettres. For them science provided the basis of a new culture. In all of these cases it was the uses of science which were most clearly visible and thus most widely known among the public.

The dissemination of science as a way of knowing was further spread among the public at large. Sir Humphrey Davy caught this new mood at the turn of the century when, in designing his plan for improving the Royal Institution in London, he put the case very directly. He hoped that the practical worker could benefit from being instructed as to the correct scientific theories of his particular branch of labor and that, in turn, he would freely communicate to the philosophical inquirer the nature of his methods so that they might be corrected by scientific principles. This represents a clear attempt to make science directly relevant to new industrial pursuits. From this diffusion of knowledge and its utilization, whatever power that science exercised during the course of the late eighteenth and early nineteenth centuries was derived. This dissemination of a way of knowing did of course represent a form of politicization and opened the way for potential external determination of the shape and actions of science.

During the course of the nineteenth century we witness what may be characterized as struggles for turf. This was the period of the establishment of the professional characteristics of science. It was also a period of reestablishing cognitive boundaries. During the nineteenth century, particularly its early decades, every one of the specific disciplines, from anatomy and astronomy to statistics and zoology, established a disciplinary identity. The A to Z of the sciences was set out in its cognitive framework, with new disciplines defining the problem areas within which scientists would work. In addition, each discipline formed its own professional society, and all the disciplines campaigned for the establishment of new specialized university professorships and institutes.

This was also a period of working out the social boundaries of science. Who would be included and who would be excluded? Closely connected with the professional move, there was a move to reset science in a social locus within society, which would give it position, status, and ultimately power. The heightened pace of the Industrial Revolution enhanced the recognition of the utility of scientific knowledge.

From the turn of the century it became clear that science would play a basic role in the new industries. This was the message that emerged from the founding of the Ecole Polytechnique in Paris in the closing years of the French Revolution, as science was directly linked to technical, industrial, and military pursuits. Similar was the message that in England motivated the establishment of the Royal College of Chemistry and the Royal School of Mines in the early decades of the nineteenth century. Both were directed toward the practical applications of science and both tied education in the sciences to industrial development. The scientists of this new generation questioned and explored what role they might play. In 1853, for example, George Hopkins, in his presidential address to the recently founded British Association for the Advancement of Science, put the utilitarian claim very directly: "One great duty we owe to the public is to encourage the application of science to the practical purposes of life, to bring, as it were, the study and the laboratory into juxtaposition with the workshop." The image being created was of the laboratory side by side with the workshop.

But even as scientists provided industrial power, and gained new status with it, they attempted to obscure the very interests which were responsible for their position and status. It is at this point in history that we see the beginning of an internal tension which has remained with science right up to the present day. Some scientists attempted to back away from the implications of the utility which had become obvious. And often it was a utility which was influential in the cognitive processes in which they were engaged. These interests became more noticeable in the prolonged and intense debates and controversies which took place in science during the nineteenth century. Richard Owen, for example, recognized what lay beneath the surface in the French debate between Pasteur and Pouchet and the earlier debate between Cuvier and Lamarck and Geoffrey St. Hilare. He put the matter very bluntly: "Pasteur, like Cuvier, had the advantage of subserving the prepossessions of the party of order and the needs of theology." Power was gained through serving established authority in church and state. Science's position was improving, its power increasing, and its ties to the state becoming clearer.

During this period of explicit attempts to utilize science in new industries, such as synthetic organic chemicals and electricity, there was also an explicit move to distinguish "pure" science from "applied" science. The very call on the word pure and the intimation of purity suggests by implication, of course, that the non-pure is somehow contaminated. There was certainly an attempt to give some special sanction and special legitimacy to the newly defined pure science. Prince Albert, one of the great patrons of science in mid-century Britain, set the tone for describing the "pure" scientists' tasks. He said they should accept "a self-conscious abnegation for the purpose of protecting the purity and the simplicity of their sacred task." The words are clear — something very special, something very strong, something separated from other activities. Lyon Playfair, an industrial chemist who himself had gone to Parliament and then to university, put the issue equally clearly. What is it that science should look like? Who should be rewarded? "The discoverer of abstract laws," he wrote, "is the real benefit of his kind — far more than he who applies them directly to industry." The aim of science? Finding new knowledge, he said. The true scientists, he claimed, were men who are "looking for sublime truths, careless of whether they will have any immediate effect on industry."

It is just this vision of the separation of the utility from the knowledge itself which came under fire from Jurgen Habermas: "Because science must secure the objectivity of its statements against the pressures and seduction of particular interests, it deludes itself about the fundamental interest to which it owes not only its impetus, but the conditions of possible objectivity themselves." It is curious, then, that even as science became industrially vital, became an important part of the productive forces of society in the nineteenth century, it attempted to "denormatize" itself, to take away from itself any vestige of interest in utility or those values which might guide it. In a sense, science seemed to want to cleanse itself of interests, even of narrow professional interests.

William Whewell, the mathematician-philosopher, put it very bluntly. "Knowledge is power," he wrote in mid-century. "But," he added, "for us it is to be dealt with as the power of interpreting nature and using her forces, not as the power of exciting the feeling of mankind and providing remedies for social evils on matters where the wisest men have doubted and differed." Hence the separation. Somehow knowledge was to be pulled away from the very power which it was gaining. The point, of course, is that power itself and the interests involved with it exerted a profound influence on the cognitive patterns and choices that scientists were making. And attempts were being made to push these very things aside. Both the professional interests and the broader social interests — interests of theology, the state, and industry — were guiding science, shaping it, and molding it in those years. Science itself was indeed being shaped by power and its place in society was being affected this way — the things it could do, the tasks it would undertake, its own visions of what it should do for its students, the novices, and its practitioners were being shaped by the new power which it was gaining. But, even as science gains power, the process is not unidirectional. The interests which affect it are not a nuisance; they are not an intrusion into an otherwise pure and rational procedure. Even though we want to avoid sociological reductionism, I think we must state the clear case of this tension which came to exist as knowledge gained and used power, became powerful itself — the tension that came to exist between science on the one hand and society on the other, between the cognitive realms and the social realms of the activity.

Where should science be practiced? What should its aims look like during the course of this newly developed strength that it had? Curiously enough, to the historian, science made a beeline for the universities and decided that it was there, in the ivy-covered towers, that science should, itself, become an established base of power. From the seventeenth century on, it had not really been an important force in the universities. In fact, most science was carried on outside the walls of universities during those two formative centuries. And it is only in the course of the nineteenth century, and with great pressure, that science managed to enter the universities and secure its power-base there. And we know that its role in the universities was greatly resented by the traditional university inhabitants. One of these, Charles Lutwidge Dodgson, better known as Lewis Carroll, wrote a brilliant and critical letter to the *Pall Mall Gazette* in 1872, in which he shows the kind of resentment that the traditional member of the university (Dodgson himself was a mathematician) had. Let me cite some parts of it just for its wit and its wisdom:

> Let me sketch in dramatic fashion the history of science's recent career at Oxford. In the dark ages of our University, some five and twenty years ago, while we still believed in classics and mathematics as constituting a liberal education, natural science sat weeping at our gates. 'Ah, let me in', she moaned, 'Why cram reluctant youth with your unsatisfying law? Are they not hungering for bones, ye panting for sulphurated hydrogen?' We heard and we pitied. We let her in and we housed her royally. We adorned her palace with reagents and retorts and we made it a very charnel house of bones. And we cried out to our undergraduates, 'The feast of science is spread. Eat, drink and be happy.' But they would not. They fingered the bones and they thought them

> dry. They sniffed at the hydrogen and turned away. Yet for all that science ceased not to cry, more gold, more gold. And her three fair daughters — chemistry, biology and physics, for the modern horse leach is more prolific than in the days of Solomon — cease not to plead give, give. And we gave. We poured forth our wealth like water. I beg your pardon, like H_2O. And we could not help thinking there was something weird and uncanny in the ghoul-like facility with which she absorbed it.

Lewis Carroll (Charles Dodgson) clearly mocked science as it entered the universities. But, in spite of his opposition and that of other traditionalists, enter science did. We can observe an almost dialectical process occurring. For even as science gained in utility to industry, it took flight into the universities. We can watch the establishment of that tension between the professional interests on the one hand and the industrial interests on the other. At no time, though, were these ever totally separated. We can watch, as science moved into the universities, the institutionalization of knowledge as power, and we can note the transformations that occurred in institutional forms — the technische Hochschulen established in the German-speaking world, the technical institutes of the English-speaking world, the polytechnics of the French world. But these professional interests were set against a hierarchy, a hierarchy with pure scientists on top, scientists located within the universities, not in the institutes. And a second tension, an institutional tension, can be seen developing. And we know that this very separation and hierarchy appear to this very day. The pure scientists seem to be those who are the definers of science and, indeed, when it comes to serving as advisors to governments, more often than not it is the pure scientists, the physicists, and not the engineers, who are chosen.

In viewing science in the universities, we can adopt an almost archeological approach. Wander around one of the older universities. The new buildings are the ones inhabited by scientists. In the older universities these generally form a ring around the central core. They are larger than the buildings of other faculties and their location demarcates them. In the universities new curricula were developed, not only for the sciences, but new curricula which involved science for many other people. The sciences, as they moved into the universities, developed an emphasis on research, replacing the moral training or the tradition-building that had often been the role of the universities. And we can watch new institutions being formed, often on the periphery of the universities, and later moving into and reforming the universities themselves.

The nineteenth century was also a period of professionalization. And, having written about this topic at great length at an earlier time, I won't review those notions here except to indicate that this pattern of the professionalized science — the driving-out of the amateur, the declaring of new roles and new ways in which scientists should act — developed and spread through Europe to the United States and, in fact, to all parts of the scientific world. What did it do? It enormously strengthened science organizationally. For science, having entered the universities, gained the ability to train the new generation, to set the norms and the problems, and the patterns by which the new generation should act. And as such, science — institutionally and organizationally — became professionalized, controlling what the activity should look like and what the novices and recruits to the field should learn and do. We can see in this professionalization the beginning of the process of making organized science a powerful social institution as well as a provider of power for other institutions and societies. But, remember the words of George Bernard Shaw. "The professions," he proclaimed, "are a grand conspiracy against the laity." Institutional and professional power, therefore, were clearly to be gained at social cost. For the laity indeed saw a separation occurring. What scientists applauded — this increased differentiation, this new status, the reduced amateurism — all of these were seen as a threat by other segments of the public. To have power, indeed, is to become suspect. Remember Acton's words about what it is that power does to you.

And as science gained its power, it also gained its critics. Take a quick look for a moment at the vivisection controversy that developed in the closing decades of the nineteenth century. I have been engaged in a restudy of it recently, and a number of interesting elements emerge: the clear professional interest being stated; the desire to do experiments with live animals as a means of bettering the position of the physiologists, of giving them more status and, indeed, of advancing their knowledge. But experiments with live animals ran counter to broadly held public mores, mores relating to the humane treatment of animals. There are also other elements in this vivisection debate which reveal some of the political problems that science was going to inherit. Strong among the anti-vivisectionists were women. One of the leaders was Frances Power Cobb, who was not only an anti-vivisectionist, but also a vigorous suffragette, a feminist. She, and others like her who joined the anti-vivisection movement, did so because of their distrust of physicians, physicians who, they pointed out, were always male, physicians who, they felt, mistreated women and didn't understand women's diseases. The anti-vivisection movement also gained adherence from others who were opposed to the institutionalized power of science as they saw it in matters such as compulsory vaccination. They objected to this compulsion to act exerted by experts. They saw a scientific establishment, and they wished to challenge this scientific establishment.

This is the period when we developed the pattern of special knowledge being held by special people. The nineteenth century was witness to the invention of whole new departments in universities and whole new institutes. Ever higher thresholds were established for entry into these fields. This is part of the meaning of professionalization. There is a pattern of exclusivity and inaccessibility. Knowledge is to be gained and used by special people, and it is a very special knowledge. You have to know too much to get to it. Note the concomitant fear of separation and loss of public legitimation that came at just this time. This is the period when we can see the movement toward popularization in the sciences; the days of Thomas Henry Huxley's lectures to the working man which he early abandoned; the period of John Tyndall, of Emil du Bois-Reymond in Germany, and numerous others who attempted to bring science to the public.

There was an ambivalence here. Should the separation really occur? And if it did, what would the consequences be? If you were to avoid the separation, how do you go about it? Well, we watched them establish the mechanics institutes which had as their aim the bringing of the knowledge of the sciences and applied sciences to those who needed technical training. But what came with them? Also a socialization to technical norms so that the skilled workers could accept the patterns being developed by the experts. This was the period of development of public or semipublic institutions aimed in some way at bringing the sciences to different groups — the Royal Institution, for example, which was established at the turn of the century, with its exhibits and public lectures, was initially aimed at the mechanics and the skilled artisans, but finally ended up providing lectures to white-tie audiences of London's upper classes and social elite. This was the period when in France, for example, museums and exhibitions — the famous Conservatoire des Arts et Métiers in Paris and the numerous copies of it in other French centers — were established, where machines and technical matters, the technical activities of the artisans and the merchants, were brought forward. These, however, were largely unidirectional efforts, that is, they were efforts on the part of the technically trained — the scientists and the engineeers — to bring their material, their views, their outlook to the public. The other part of the feedback loop was not closed. The public had very few ways of speaking directly to the scientists or the experts themselves. Why do I dwell so long on these seeming factors of separation and antagonism? In part, because of what was happening. We can watch the slow creation not only of a knowledge elite — indeed, science was becoming more complex and harder — but, alongside the knowledge elite, the creation of a social elite. Mind you, scientists were not to be found at the highest points of social power. Think, for example, of our national leadership in relation to knowledge and intellect. Scientists are rarely among them. But, the scientists had become, as C. Wright Mills, the American sociologist, some years ago called them, the technical lieutenants of power, the people just below the top level who would exercise power in concert with the leaders of the state.

The twentieth century represents an era of experts and their powers, and additionally their problems. By the end of the nineteenth century, modern states had come to recognize the need for the sciences, not as sharpiy delineated, but certainly seen as applied sciences. They were willing to tolerate the pure, but they certainly wanted the applied. By mid-century, Prince Albert had already discerned a kind of tie and relationship which was developing. In his presidential address to the British Association in 1859, he foresaw the day when science "will no longer require the begging box, but will speak to the state like a favoured child, sure of his paternal solicitude for his welfare." Indeed, the position of science changed. Getting the patronage of the state came slowly, but come it did. Some of the earliest patronage, and indeed some of the earliest influence, however, came through the new industries, including industries which provided such basic things as a new source of power in electricity. Here you had an industry dependent upon, and therefore closely linked to, the sciences. In the public mind, of course, this link always existed. The link between the newly developed technologies and industries and the sciences loomed large. Whereas scientists might not feel the closeness of this tie, popular literature, popular illustrations, popular books of the day showed it over and over again. This, of course, was a form of legitimation The public saw science as giving it something it wanted, as well as providing national strength. But, of course, just this form of legitimation — that is, giving sanction to science because of its productive abilities — worried some of the insider scientists. They were afraid of the ultimate feedback. Power could be seen in the products and in the potentials of science. And by the year 1900 there was a scientism which spread broadly over most of the Western societies. All things were knowable and all problems solvable if only science could be put to work. The interests of the scientists, then, became identified, from both within and without, with industrial success. And scientism became part of the outlook of the modern industrial societies.

But this can give rise to problems: When industry fails, as during the great depression of the 1930s, science can be seen as being at fault. And during the 1930s, in Britain and in the United States and on the Continent, we heard calls for moratoria on the gaining of new knowledge and technique. There was enough to be had, the critics said. Stop with this new knowledge; stop with the new technique. Let's learn to use what we have; let's set our house in order before we go off after further innovations. In a peculiar way there was a mirror image of this attitude within science. It said, let us transform the practice of science to aid the creation of a new social order — the social order of socialism itself. And we noted the growth, particularly among a group of British scientists in the 1930s, of a movement to use science in this socially progressive manner. One critic suggested that they ought to do away with the House of Lords and establish a senate of scientists to deal directly with the real problems in society. There was a sizeable amount of internal dissension within the sciences at the time, and there was a conscious clash of interests among scientists — some adhering firmly to the parties in power, others indeed aligning themselves with the radical critics. Here we see normative views emerging explicitly, that is, social images and social values compelling scientists to redefine what it was that they felt science could do and should do. This whole debate during the 1930s, which is a fascinating one, recently spelled out in a book by a former student of mine, Gary Wersky, can be set and seen against an interesting background — the idealized image of science which the protagonists saw as being at the core of society and the state in the Soviet Union. Think of the implications, then, for the sciences when the Soviet Union became a focus of criticism later on. We can watch, even there in the 1930s, the beginning of a process that I have called de-institutionalization and de-professionalization, that is, the breakdown of consensus within the scientific community and the breakdown of the organized patterns of science itself. People were willing to step outside the professional boundaries to become critics of the material in which science was involved and the activities in which it was engaged. Incidentally, I might point out that I can understand some of the anger that people expressed at the works of an historian and sociologist of science like Thomas Kuhn when he compared normal science to puzzle-solving. Because, remember, it was during the 1930s that J.D. Bernal, the British communist scientist, compared basic science to the solving of crossword puzzles.

The attitudes that developed toward science, both from within the scientific community and from without, during the years of the depression are harbingers of what was to come. I suspect, though, that the major turning point can be seen in the decision to produce and use the atomic bomb. Samuel K. Allison, one of the American physicists directly involved in the project, came away from the making of that weapon with the rather dire conclusion that physics and physicists are important for waging war. And

Edwin U. Condon, another of the key figures engaged in the activity, wondered whether the revulsion to war that was to come after the use of the bomb might extend to the sciences. And, indeed, it is exactly this revulsion to war, and the link of the sciences to war, which we could see in the Vietnam decade and the attitudes many young people had toward science and scientists in the universities. Science, through its involvement in the bomb, had become politically interesting. And we know that because of this involvement, scientists themselves became interested in politics. A question was raised, a question about the whole culture of science as it developed in the postwar years. Oppenheimer talked about the problem, as I indicated in the opening sentences of this paper. But there were other things in Oppenheimer's career, and in the careers of many scientists working on the bomb, which came out only incidentally as we probed more deeply. What about the scientists at the time they were actually engaged in producing the bomb? Were there no critics? Well, there were some, we know. But, as Oppenheimer put it in his testimony, they went ahead with the project, not asking the moral questions about its potential use, because the problem was "technically sweet," to use his term. It is just this view that Freeman Dyson, a mathematical physicist at the Institute for Advanced Study at Princeton, captured in an article in *The New Yorker*, later turned into a book. Nuclear explosions, he said, have a glitter more seductive than gold to those who play with them. To command nature, to release in a pinpoint the energy that fuels the stars, to lift by pure thought a million tons of rock into the sky — those are exercises of the human will which produce an illusion of unlimited power. Power, then, given to the scientists. And what happens when you have unlimited power? Dyson probed further into the roles of Oppenheimer and of Edward Teller, the man who so strongly backed the making of the hydrogen bomb. He recognized the contrast and conflict between these two, and then he analyzed what it was that they were engaged in. He said:

> But each of them, having achieved his technical objective, wanted more. Each of them was led by his Demon to seek political as well as technial power. Each of them became convinced that he must have political power to ensure that the direction of the enterprise he had created should not fall into hands that he considered irresponsible. In the end, each of them was irrevocably committed to exercises of the human will in the political as well as the technical sphere. And so each of them in his own way came to grief.

What is it that power does? Indeed, some level of that use of power and its availability in the current scientific enterprise leads to some sort of corruption.

There are also, however, other successes we can watch; successes of a peculiar sort which have changed the nature of the relationship between the experts and the lay person. What were they? The very successes in the education in the sciences given to the public. In those years that we all spent in schools learning sciences and mathematics, things that our parents never had, we gained the potential for the demystification of the sciences. We gained the potential for being less afraid of questioning the experts.

There were also, of course, deeper political trends which led to this kind of questioning of the experts. They were the politics of participation that swept Europe and the United States — in the U.S. in the Civil Rights movement and the anti-Vietnam War movement; on the Continent, the breakdown of the authoritative relationis that had been prevailing since the end of World War II. There was a breakdown of consensus at many, many levels — a breakdown certainly at the political level, but there was also a breakdown of consensus within science itself. There were attacks on the sources of power and on science as elites. And science was challenged, both as a source of power and as the domain of a social elite. Its candor and its honesty were questioned; its interests and its ties were explored. Indeed, one of the men who gave the clearest commentary and set the clearest guide for this critical exploration was himself a great military leader, the late President of the United States, Dwight Eisenhower. Remember his talk about the military-industrial complex, and then his talk about the military-scientific complex? He was aware of just what it was that was happening, and he gave focus to a relationship between industry, experts, and the military and what might come of it. The attacks, of course, didn't stop at science as a social and political force, but they went further. During the 1960s and the early '70s they challenged the very epistemology, the very way of knowing of science itself, especially science's inability and failure to deal with normative or value issues. The claim of being value neutral or value free was scoffed at, as science was seen to use its power in negative ways. Where were the values?, the critics asked. Which values were being followed? Which interests were in command? Some of these challenges actually came from practitioners of science themselves, and some, indeed, from the science watchers — historians, philosophers, sociologists, and political scientists who were studying science itself. In recent days, we have been witnessing an attempt to understand and reconstruct the activity that we call science. And there is a great deal of excitement in all of these fields — history, philosophy, sociology, and political study of science. All the major issues seem open. The critical questions are now being asked and authority is no longer unchallenged. My own hope is that the philosophy of science, the social study of science, can share in these probings and that we must not attempt too quickly to force limited understandings or closures of debates, closures that don't really exist. After all, if there is one thing we have learned as we have watched the climb of science from the will to attain power to the exercise of power, it is that the enterprise of science itself is too powerful to be left to the experts. The role of the citizen needs full exploration.

An alternate version of this paper has appeared in: *Science Under Scrutiny, The Place of History and Philosophy of Science*, ed. Roderick W. Home, Australasian Studies in History and Philosophy of Science, 3 (Dordrecht: Reidel, 1983).

Knowledge and Power in the Sciences
A Comment

YARON EZRAHI

Everett Mendelsohn claims that since the late nineteenth century, a combination of factors has transformed scientific knowledge into a form of socially illegitimate and abusive power. According to Mendelsohn, these factors include the tendency of science to serve powerful interests rather than to be neutral; the role of science in magnifying the power of the nation-state and its involvement in modern wars and mass murder; the professionalization of science which has increasingly excluded laymen not only as co-participants in, but even as an audience of, scientific discourse. Finally, science, according to Mendelsohn, has contributed to the substitution of technical training for moral education. He echoes Rousseau's fear of specialists unrestrained by a comprehensive moral outlook. As such, he observes, the scientists become involved in what George Bernard Shaw called "a grand conspiracy against the laity."

These developments have created, according to Mendelsohn, a crisis in the legitimacy of science in the modern liberal-democratic polity. This crisis is still unresolved despite some recent steps taken to enlarge the role of the lay public in influencing the scale and directions of the scientific enterprise.

I concur with the Professor Mendelsohn's point about the corrosive effects that the social uses and misuses of science have on its social legitimation. I think, nevertheless, that his discussion suffers from an overly restricted and

E. Ullmann-Margalit (ed.), The Kaleidoscope of Science, 241–245.
© *1986 by D. Reidel Publishing Company.*

narrow conception of the relations between knowledge and power. The historical record of early as well as later developments indicates much more ambivalent relations. Science often simultaneously reinforces and undermines democratic handling of power. The same factors which amplify power often also serve to curtail and limit it. A dialectical conception of the relations of knowledge and power would, therefore, seem both more accurate and more useful for discerning the complexity of their interactions and interpenetrations. I think such a conception will help to show the extent to which, contrary to Mendelsohn's argument, science has been used also to check and not only to boost political power.

Inasmuch as Mendelsohn concentrates on the corrosive effects that knowledge as power has on democratic practices, I would like to dwell on those aspects of the dialectical relations of knowledge and power where science proved congenial for grounding democratic conceptions and practices of power.

Very briefly: Some of the basic challenges of political authority in democracy have been how to harmonize the commitment to politics as an enterprise of diverse persons with the establishment of power untainted and incorruptible by subjectivity and personal caprice; how to affirm individual freedom and equality and yet institute restraint; how to reconcile the cherished value of individualism and the necessary requirements of public order. One of the contemporary theorists of democracy has observed that "the depersonalization of power has indeed been a task to which Western man has devoted all his political ingenuity."[1] In the evolution of the modern democratic techniques for public rather than personal and hierarchical handling of power, science has played a very important role. This can be shown if we examine characteristic democratic approaches to ensuring the accountability of both political speakers and political actors. As both a scientist and a sociopolitical thinker, Condorcet saw this point clearly enough, although he formulated it in a way which makes it sound utopian to modern ears:

> In all countries where the physical sciences have been cultivated, barbarism in the moral sciences has more or less been dissipated ... the more men are enlightened, the less necessary it will be to give [to men of authority] social powers, energy and extent. Thus truth is the enemy of power, as of those who exercise it. The more it spreads the less they will be able to mislead men ....[2]

The record of modern democracies shows that these hopes, though admittedly utopian and largely unrealizable, left a powerful legacy in the practice of the liberal democratic state. With respect to the public accountability of speakers in the political sphere, the example of science reinforced the trust in the feasibility of civil discourse in which disputes are at least partly resolvable by reference to publicly accessible evidence.

Against a historical background of divisive and disruptive theological disputations. Bishop Spratt asserted the superiority of the plain language of the Fellows of the Royal Society.[3] He celebrated their rejection of ornamental and affective modes of speech and praised their mode of irenic rather than polemic discourse. The historical success of science added force to the democratic trust in language as an alternative to violence and as an instrument of restraint which is compatible with freedom. Science was an example for securing the accountability of speakers by public tests of logic and evidence. The many modern instances of abusing the scientific concept of truth by enlisting the authority of science to fix ideological dogmas qualifies but does not undercut suggestions, made in the nineteenth century by thinkers such as J.S. Mill and in the twentieth century by J. Dewey and K. Popper, that the canons of scientific discourse have close affinity to liberal-democratic principles of civil discourse. The role of scientific discourse as an example for controlling subjective and personal bias and for holding speakers publicly accountable was often compared unfavorably with the example of poetry. Poetical discourse was characteristically described as an example for the subjective, capricious, unaccountable use of language inappropriate to the context of public affairs.

In their response to the attacks of twentieth-century ideologues of totalitarian states, democratic spokesmen have characteristically stressed the features of democracy as government by discourse rather than by indoctrination. They have often referred to science as an illustration of these principles. The entire system of a free press based on the value ascribed to reports on the world independent of political institutions was nourished by the example of science.

Today the various expert communities, which Mendelsohn rightly notes are largely exclusive professional forums, are at the same time vital for inculcating responsible — publicly accountable — speech in fields of government action such as economic, health, welfare and defence policy. The statistical and social indicators which professionals employ in free societies to measure the extent of inflation, poverty, unemployment, pollution and other problems restrain political speakers in their freedom to deceive the public.

The contributions of science to the democratic accountability of actors have been equally important. Our modern preoccupation with techniques as methods of control and manipulation often obscures the constant democratic preoccupation with the need to curtail arbitrary, personal use of power. The legal code and the institutional instruments of its social application have constituted important means to ensure the public accountability of political actors. Science and technology, by carving out aspects of human actions that can be judged and assessed externally and independently of the personalities of actors, furnish complementary procedures.

When the scientific "externalization" of human conduct is criticized as dehumanizing, the fact that at the same time it has been a response to the threat of the personal exercise of power is often ignored. But basing public action on technical competence rather than on personal trust or ideological qualifications was a strategy for decentralizing power in government structures. Science and technology have thus contributed to the redefinition of the accountability of actors in terms of performance rather than in terms of ascriptive personal traits or political affiliation. Michael Faraday furnishes an early illustration of the role of the scientist in this development. In 1844 there was an explosion in Durham's coal mines, causing many casualties. The explosion generated considerable unrest in the region, making it desirable to identify the causes of the explosion quickly and authoritatively. Faraday was asked to investigate the matter and determine if it was an unavoidable accident or a case of negligence. His judgment as a scientist was crucial for the assignment of blame to, or exoneration of the parties involved.[4] This was, of course, only a characteristic early example of the increasing role of science in defining public technical criteria for the attribution of responsibility to actors and for publicly authoritative assessment of the rationality of actions. The fact that this power could be abused does not invalidate its contribution to the democratic polity any more than abuses of the law support an argument against constitutional government.

Such positive contributions of science to the democratization of power, and the public accountability of political authority do not negate, but rather coexist with, the uses of knowledge to restrict democratic processes which Professor Mendelsohn stressed. Taken together, the simultaneous uses of science to decentralize political power and to amplify it, to ground public criticism of government actions, and to help it hide its political objectives behind a façade of technical arguments, reveal the dialectical relations between science and power. The involvement of science with both democratic and counter-democratic trends complicates the problem which Professor Mendelsohn puts before us: the problem of how scientific activity can be made responsible and legitimate in the modern democratic state.

Enhancing political controls over science may reduce the potential of science as a force unaccountable to the larger society, but this strategy will surely deprive democracy not only of one of the most meaningful spheres of human endeavor, but also of some of the most effective instruments for holding decision-makers and actors publicly accountable by impersonal standards of competent performance. Democracy cannot be saved from science without also becoming less democratic. In the late twentieth-century democracies, the political logic of balanced pluralistic representation has already proved incongenial for the application of intellectual standards in public affairs. In societies where policy-making is the result of negotiation rather than of the search for "best solutions," science can neither effectively reinforce nor restrain political power.

Decades of relentless efforts to bring scientific knowledge to bear upon public policy demonstrate that in the constant encounter between scientific and technical standards of functional adequacy and considerations of political feasibility, the latter almost consistently have the upper hand.[5]

This is why I would suggest a small modification in Professor Mendelsohn's way of putting the question about the legitimation of the power of knowledge in modern liberal democratic society. I think the issue today is not so much how to increase citizens' control over the power of science. That has already been partly achieved. It is rather how citizens can hold political speakers and actors publicly accountable while the authority and technical standards for responsible political speech and action are weakening. The problem is that the politicization of science and of its authority in public affairs largely undermines the very norms which citizens in a free society need to apply in order to check the misuses of both knowledge and power.

Notes

1 Giovanni Sartori, *Democratic Theory* (Detroit: Wayne University Press, 1962), p. 408.
2 Cited in Keith M. Baker, *Condorcet: From National Philosophy to Social Mathematics* (Chicago: Chicago University Press, 1975), pp. 75, 79.
3 T. Spratt (1667), *History of the Royal Society*, ed. J. Cope and H.W. Jones (St. Louis: Washington University Press, 1958).
4 Morris Berman, *Social Change and Scientific Organization, The Royal Institution 1799–1844* (London: Heinemann, 1978).
5 Edward C. Banfield, *The Unheavenly City Revisited* (Boston: Little Brown, 1970), pp. 260–286.

Index of Names